Artan Dermaku

Path recognizing for Robots through low cost sensors

Artan Dermaku

Path recognizing for Robots through low cost sensors

From Mobile and Humanoid Robot perception model up to simulation of several heuristic approaches for path planning

Südwestdeutscher Verlag für Hochschulschriften

Impressum / Imprint
Bibliografische Information der Deutschen Nationalbibliothek: Die Deutsche Nationalbibliothek verzeichnet diese Publikation in der Deutschen Nationalbibliografie; detaillierte bibliografische Daten sind im Internet über http://dnb.d-nb.de abrufbar.
Alle in diesem Buch genannten Marken und Produktnamen unterliegen warenzeichen-, marken- oder patentrechtlichem Schutz bzw. sind Warenzeichen oder eingetragene Warenzeichen der jeweiligen Inhaber. Die Wiedergabe von Marken, Produktnamen, Gebrauchsnamen, Handelsnamen, Warenbezeichnungen u.s.w. in diesem Werk berechtigt auch ohne besondere Kennzeichnung nicht zu der Annahme, dass solche Namen im Sinne der Warenzeichen- und Markenschutzgesetzgebung als frei zu betrachten wären und daher von jedermann benutzt werden dürften.

Bibliographic information published by the Deutsche Nationalbibliothek: The Deutsche Nationalbibliothek lists this publication in the Deutsche Nationalbibliografie; detailed bibliographic data are available in the Internet at http://dnb.d-nb.de.
Any brand names and product names mentioned in this book are subject to trademark, brand or patent protection and are trademarks or registered trademarks of their respective holders. The use of brand names, product names, common names, trade names, product descriptions etc. even without a particular marking in this work is in no way to be construed to mean that such names may be regarded as unrestricted in respect of trademark and brand protection legislation and could thus be used by anyone.

Coverbild / Cover image: www.ingimage.com

Verlag / Publisher:
Südwestdeutscher Verlag für Hochschulschriften
ist ein Imprint der / is a trademark of
OmniScriptum GmbH & Co. KG
Bahnhofstraße 28, 66111 Saarbrücken, Deutschland / Germany
Email: info@omniscriptum.com

Herstellung: siehe letzte Seite /
Printed at: see last page
ISBN: 978-3-8381-3229-7

Zugl. / Approved by: Vienna, TU-Wien, Diss.,2012

Copyright © Artan Dermaku
Copyright © 2012 OmniScriptum GmbH & Co. KG
Alle Rechte vorbehalten. / All rights reserved. Saarbrücken 2012

Abstract

The aim of this work is the development of a new methods oriented on the creation of new software approaches which will try to calculate an optimal or nearly optimal path from robot to the target object. These new heuristic algorithms are based on a grid-map representing the working robot environment. For map creating, different state of the art approaches that use the data-fusion, error and uncertainty and software solutions are proposed. Considering the data obtained from multiple sensors can be decided for the new moving steps of robots on partial segment of the path, as well an optimal behavior of robot when it stay in front of target objects.

The obtained data from multiple sensors should make possible also an optimal navigation of robot in terms of avoiding collisions between robot and objects. Using of the Extended Kalman Filters and Neural Network approach should make possible the updating of the robots position during its localization. Especially the applying of Neural Network leads to the reducing of the robots position error and its uncertainty, as difference between desired and achieved position.

The new in this work implemented heuristic algorithms, are based on hypertree decomposition as well as on geometrical intersection between robot position, obstacles and goal position.

The simulation application in C# and Matlab is also implemented. This application use pre-defined work environment with obstacles as grid-map, and for different start positions, goal positions and heuristic algorithms finds and simulate the optimal or near optimal path.

The main advantages of such approaches are the time reducing of map-calculating and replacement of expensive high-tech devices with low-cost sensors and better software solutions.

For objects detecting, the vision-system (not in this work implemented), as part of perception is proposed. It makes possible the map calculation of robots work environment.

Finally, simulations in Matlab for real robots HUMI and ARCHIE implemented at the Institute of Handling Robotics and Technology (IHRT), are done. Fuzzy Logic Controller for HUMI as well as PID Controller for Archie are implemented and tested.

Dedication

*This thesis is dedicated to my lovely daughter **Ana**, who is the inspiration of my life.*

Acknowledgments

Hereby I would like to thank my supervisor, em.o.Univ.Prof. Dr.techn.Dr.h.c.mult.Peter Kopacek , for his patience with me and for his willingness to help me any time and to answer me in any question concerning this PhD work. I will be forever grateful.

In addition, I would like to thank Professor Dr.techn. Numan Durakbasa, for the proofreading of this PhD work, and for very useful suggestions.

I would also like to thank and appreciate Msc.Xhevahir Bajrami from Mechanical and Mechatronic Institute, for his great help and advice in mechanical and kinematical issues during the PhD work, as well as for great time spending at the institute. I would like to thank Mr.Peter Unterkreuter for interesting discussions.

Also, I want to thank Mr. Hakif Gashi, Mr.Rabit Halili and Mr.Skender Blakqori for supporting me during my studies.

I thank also "Iliria" University, especially the president Prof.Dr.Mixhait Reçi, for the understanding for my absence during the last months.

My most profound thanks are to my mother, Fahrije Dermaku for her infinite love and devotion, my father Professor Ejup Dermaku who was the inspiration for my success, to my sisters Vetiola and Shqipe and my brother Besjan, for their love and understanding, support and trust that they all showed me all my life.

Finally, for her love to me, for her unconditionally understanding during my absence, for her support in every step I made, I will be forever grateful to my wife Msc. Krenare Sogojeva Dermaku. Thank you.

Contents

Problem Formulation .... -1-
1   Introduction .... - 5 -
   1.1   Autonomous Mobile Robots .... - 5 -
   1.2   Humanoid Robots .... - 8 -
   1.3   Navigation and Path planning .... - 13 -
2   State of the art .... - 17 -
   2.1   Perception .... - 19 -
      2.1.1   Sensors in Robotics (Siegwart and Nourbakhsh, 2004) .... - 19 -
         2.1.1.1   Tactile sensors .... - 20 -
         2.1.1.2   Wheel motor sensors .... - 20 -
         2.1.1.3   Motion/speed sensors .... - 20 -
         2.1.1.4   Vision-based sensors .... - 20 -
         2.1.1.5   CCD technology .... - 21 -
         2.1.1.6   CMOS technology .... - 22 -
         2.1.1.7   Visual ranging sensors .... - 23 -
         2.1.1.8   Stereo vision .... - 23 -
         2.1.1.9   Color-tracking sensors .... - 24 -
         2.1.1.10 CMUcam robotic vision sensor .... - 24 -
         2.1.1.11 Active ranging sensors .... - 25 -
         2.1.1.12 The ultrasonic sensor (time-of-flight, sound) .... - 25 -
         2.1.1.13 Laser rangefinder (time-of-flight, electromagnetic) .... - 26 -
         2.1.1.14 Ground based beacons .... - 26 -
         2.1.1.15 The global positioning system (GPS) .... - 27 -
      2.1.2   Representing of errors and Uncertainty (Siegwart and Nourbakhsh, 2004) .. - 28 -
         2.1.2.1   Gaussian Distribution .... - 30 -
   2.2   Localization .... - 31 -
      2.2.1   Noise and Aliasing Problem (Siegwart and Nourbakhsh, 2004) ... - 32 -
      2.2.2   Belief Representations (Siegwart and Nourbakhsh, 2004) .... - 33 -
      2.2.3   Map Representation .... - 35 -
      2.2.4   Probabilistic localization .... - 37 -
      2.2.5   Autonomous Map Building .... - 40 -
   2.3   Path Planning and Navigation .... - 41 -

2.3.1 Road Map Approaches (Siegwart and Nourbakhsh, 2004). .... - 43 -
2.4 Obstacle Avoidance – BUG Algorithm .... - 45 -
3 A new cost oriented method .... - 47 -
3.1 Perception and Vision System .... - 48 -
3.2 Map calculating of the environment .... - 52 -
3.3 Localization of Mobile Robot .... - 68 -
3.3.1 New Back-Propagation Neural Network approach .... - 76 -
3.3.2 Summary .... - 86 -
3.4 Path Planning and Navigation .... - 87 -
3.4.1 A new heuristic algor. for path planning based on geometrical intersection .. - 91 -
3.4.1.1 Summary .... - 109 -
3.4.2 A new heuristic algor. for path planning based on hypertre decomposition - 110 -
3.4.2.1 Summary .... - 129 -
4 Hardware Implementation .... - 130 -
4.1 HUMI – A Demining Robot .... - 131 -
4.1.1 Analysis of HUMI-s kinematic model .... - 131 -
4.1.2 Analysis of the Dynamic Model of HUMI .... - 139 -
4.1.3 HUMI-s new controller design in FLC (Fuzzy Logic Control) .... - 142 -
4.2 Humanoid Robot Archie .... - 146 -
4.2.1 Kinematic model of biped robot Archie .... - 147 -
4.2.2 Equations of Motion .... - 149 -
4.2.3 Dynamic modeling in SSP model .... - 150 -
5 Software Implementation and Simulation Results .... - 154 -
5.1 Map-Creation .... - 155 -
5.2 Path-Simulating .... - 156 -
5.3 Path Simulation for HUMI .... - 159 -
6 Summary and Outlook .... - 166 -
6.1 Future Work .... - 167 -
Bibliography .... - 169 -

Table of Figures

Fig. 1-1 The mobile Robot sojourner used on the Mars mission in summer 1997 (Internet) ... - 6 -
Fig. 1-2 HUMI (Kopacek , 2010) .... - 7 -
Fig. 1-3 Mobile Mini Robot .... - 8 -
Fig. 1-4 Humanoid Robot – QRIO .... - 10 -
Fig. 1-5 Humanoid Robot - Archie .... - 11 -
Fig. 1-6 Humanoid Toy Robot – Robonova .... - 12 -
Fig. 1-7 Coarse-fine planning for a mobile robot (Kensuke et al., 2010) .... - 15 -
Fig. 2-1 Navigation of mobile robotics (Siegwart,R. and Nourbakhsh.I.R., 2004) ...... - 17 -
Fig. 2-2 Cheap CCD camera (http://www.videologyinc.com/cameras/…) .... - 21 -
Fig. 2-3 Low-cost CMOS camera .... - 23 -
Fig. 2-4 Color markers on the robot and color-tracking sensor .... - 24 -
Fig. 2-5 Calculation of positions based on GPS. .... - 28 -
Fig. 2-6 Probability density function (Siegwart and Nourbakhsh, 2004) .... - 29 -
Fig. 2-7 Error propagation in system with n-inputs and m-outputs .... - 31 -
Fig. 2-8 General schema for Mobile Robot Localization .... - 32 -
Fig. 2-9 Concept of sensor data fusion for target tracking (Raol, 2010) .... - 33 -
Fig. 2-10 Exact Cell Decomposition (Siegwart and Nourbakhsh, 2004) .... - 36 -
Fig. 2-11 a) Fixed Decomposition   b) Narrow passage disappears   (Siegwart and Nourbakhsh, 2004) .... - 36 -
Fig. 2-12 "Quadtree" Cell decomposition (Siegwart and Nourbakhsh, 2004) .... - 37 -
Fig. 2-13 Localization using the Extended Kalman Filter .... - 39 -
Fig. 2-14 Visibility Graph (Siegwart and Nourbakhsh, 2004). .... - 44 -
Fig. 2-15 Voronoi Diagram (Siegwart and Nourbakhsh, 2004). .... - 45 -
Fig. 2-16 a) BUG 1 Obstacle Avoidance Algorithm   b) BUG 2 Obstacle Avoidance Algorithm .... - 46 -
Fig. 3-1 Software development concept in general .... - 48 -
Fig. 3-2 Sensor Fusion System .... - 50 -
Fig. 3-3 Image projection from 3D into 2D pixel coordinates .... - 51 -
Fig. 3-4 Perception system in general .... - 52 -
Fig. 3-5 Fixed Decomposition of the space .... - 54 -
Fig. 3-6 Occupancy grid .... - 54 -
Fig. 3-7 Cell of the map-grid in pixel frame coordinate-system .... - 55 -
Fig. 3-8 Cell-indexes and the Cell-coordinates .... - 56 -
Fig. 3-9 Binary Image Matrix .... - 56 -
Fig. 3-10 a) Cell-Grid and all its neighbors   b) Directly-connected cells: .i.e **celli** and **cellj** are directly connected, but no-one of them i directly connected with **cellk** .... - 57 -
Fig. 3-11 **celli = x_i , y_i , F_i** and **cellp = x_p , y_p , F_p** are continually-connected .... - 58 -
Fig. 3-12 Two different form of objects (map_object1 and map_object2) found .... - 60 -
Fig. 3-13 Black colored pixels show the boundary cells (pixels) of two different objects ...... - 61 -
Fig. 3-14 Two Homogeneous Objects .... - 63 -
Fig. 3-15 a) Mapping with two objects creates from CCD camera and   b) The similar mapping of the same objects created from sonar sensors .... - 64 -
Fig. 3-16 New mapping object as result of fusion of objects from .... - 65 -

Fig. 3-17 The boundary cells of mapping-object3 after certain are of time. .... - 66 -
Fig. 3-18 Block - Diagram of COR localization processes .... - 71 -
Fig. 3-19 Second pixel frame created .... - 72 -
Fig. 3-20 World coordinates and robot coordinates .... - 73 -
Fig. 3-21 The supposed calculated Trajectory of COR and its estimated positions for different time-steps .... - 77 -
Fig. 3-22 Difference between COR estimated position and position calculated on map .... - 78 -
Fig. 3-23 Neuronal schema .... - 79 -
Fig. 3-24 Activation Function and Output calculation Example for Hidden Neuron N5 .... - 80 -
Fig. 3-25 Continuous Log-Sigmoid Function [Matlab] .... - 81 -
Fig. 3-26 Example of the finding of the output-error for a hidden neuron N5 .... - 82 -
Fig. 3-27 Overview of the Back-Propagation Algorithm .... - 83 -
Fig. 3-28 Binary Occupancy grid map .... - 92 -
Fig. 3-29 Some of possible positions between COR, Target Positions and Concave Object . - 93 -
Fig. 3-30 Decision for left or right avoiding .... - 95 -
Fig. 3-31 The left and right between-cells during avoiding of the Object .... - 96 -
Fig. 3-32 Not directly-connected neighbour cells .... - 101 -
Fig. 3-33 The finding of "way-out" of an object .... - 102 -
Fig. 3-34 The identification of the Homogenous Objects and the map after creation of Homogenous Objects. .... - 106 -
Fig. 3-35 Subsets as (between-positions) H_1 to H_n between Start and Target positions.- 106 -
Fig. 3-36 Interpolation Line between SP and TP and its intersection with First Object occurred . - 107 -
Fig. 3-37 A simple Hypergraph example (a), it's primal (b) and dual (c) graph .... - 111 -
Fig. 3-38 A Hypergraph example (a) and its Hypertree Decomposition of width 2 (b) .... - 114 -
Fig. 3-39 Binary Occupancy grid map (see chapter 3.2) .... - 116 -
Fig. 3-40 Numbering of the empty cells of the grid-map .... - 120 -
Fig. 3-41 Example of a Hypertree Decomposition with its Levels .... - 123 -
Fig. 3-43 Path calculated from Algorithm based on Hypertree Decomposition for cells between number 53 and 58 .... - 128 -
Fig. 3-42 Hypertree Decomposition for Hypergraph given in Fig.3-42 .... - 128 -
Fig. 4-1 HUMI .... - 131 -
Fig. 4-2 Positional analysis and kinematic model for HUMI .... - 132 -
Fig. 4-3 MATLAB Model-solution for inverse kinematic of HUMI .... - 135 -
Fig. 4-4 Desired Trajectory of HUMI .... - 136 -
Fig. 4-5 Coordinates of point B in X axe-direction .... - 136 -
Fig. 4-6 Coordinates of point B in Y axe-direction .... - 137 -
Fig. 4-7 Velocity v_x of point B .... - 137 -
Fig. 4-8 Velocity v_y of point B .... - 138 -
Fig. 4-9: The angle θ between HUMI and X - axe .... - 138 -
Fig. 4-10 Steering angle ϕ of front wheels of HUMI .... - 139 -
Fig. 4-11 The nominal driving force acting on the rear axle .... - 141 -
Fig. 4-12 The torque for steering wheels. .... - 142 -
Fig. 4-13 The Fuzzy rules between Inputs and Outputs .... - 144 -
Fig. 4-14 Error of front wheels steering angle after applying Fuzzy .... - 145 -

Fig. 4-15 Error for angle theta between HUMI platform and X-axe after applying Fuzzy... - 145 -
Fig. 4-16 Error of front wheels steering angle without applying Fuzzy logic .... - 145 -
Fig. 4-17 Error for angle theta between HUMI platform and X axe without aplying Fuzzy - 146 -
Fig. 4-18 Humanoid Robot Archie at IHRT .... - 147 -
Fig. 4-19 a) Seven link biped model of Archie   b) Half cycle of walking in sagittal plane - 148 -
Fig. 4-20 Dynamic walking of Archie .... - 150 -
Fig. 4-21 Simulation results after applying of PID for Archie .... - 153 -
Fig. 5-1 Simulation of Map Creation .... - 155 -
Fig. 5-2 Main Window of Simulation application .... - 156 -
Fig. 5-3 Simulating of Hypertree Decomposition .... - 157 -
Fig. 5-4 Statistics of all approache .... - 157 -
Fig. 5-5 MATLAB Model solution for HUMI .... - 160 -
Fig. 5-6 Subsystem for Inverse Kinematic – M.File .... - 160 -
Fig. 5-7 Inverse Kinemtaic of HUMI - other data .... - 161 -
Fig. 5-8 Forward Kinematic .... - 161 -
Fig. 5-9 Fuzzy Logic .... - 162 -
Fig. 5-10 Errors of position and angle for HUMI-platform .... - 162 -
Fig. 5-11 Error of front wheels steering angle and velocity for HUMI-platform .... - 163 -
Fig. 5-12 Errors of position and angle for HUMI-platform with FLC .... - 164 -
Fig. 5-13 Error of front wheels steering angle and velocity for HUMI-platform with FLC . - 165 -

Problem Formulation

Navigation of mobile robots is a complex process including perception, localization, path- planning and motion control. Each of those tasks should be completely understand and from robot correct executed in order to navigate successfully.

Understanding and interpreting of real working environment by robots is done through perception process. For this purpose, data obtained from multiple sensors (for example light sensors, touch sensors, ultrasonic sensors, infrared sensors, visual sensors etc.) should be correctly extracted and interpreted by the robot, especially using the sensor fusion.

Once the working environment is correct interpreted, the robot should correctly localize himself. The building of map representing correctly and fairly real environment is the first step. The main task, by the localizing of robot, is its estimated position. Due to sensors noise and other uncertainty there can happen errors during the localizing. To minimize these errors and to find the "correct" positions of the robot on map, the different approaches like Kalman Filter (KF), Extended Kalman Filter (EKF), as well as artificial intelligent approaches like Neural Network, Fuzzy Logic etc. can be applied.

Before a robot moves from any start position to goal position, a path should be planned. This planning of path should include not only obstacle avoiding but also the optimality. Different algorithms for this purpose exist, from road-maps up to heuristic and genetic algorithms.

To follow the trajectory, the robot should be able to command and to control drives. This is done in Motion control process.

In this work, a new cost oriented method is proposed, which take in consideration Localization and Path Planning. The main advantages of such approaches are the time reducing of map-calculating and replacement of expensive high-tech devices with low-cost sensors and better software solutions.

The main parts of new proposed software development concept are:

- Map calculation of the environment using Vision System as part of perception that is able to detect and recognize objects;
- Several new approaches for optimal or nearly optimal path planning and robot navigation using heuristic and exact algorithms, neuronal network and Extenden Kalman Filters ;
- Simulations application in C# and Matlab ;

In this work a vision system consisting of CCD Camera or USB Camera and Ultrasonic Sensors is proposed. For this it is proposed OpenCV Library (Open source computer vision library in C/C++) to be used as tool for basic image processing approaches, objects recognition and motion segmentation. There is no exact object recognition required, only a frame with exact dimensions should be calculated which make possible for robot to avoid this obstacle.

The vision system obtained in this way is a basis for path planning and robot navigation. We have to differentiate between static objects and objects that move during navigation of robot. It is possible to recalculate exact positions of objects in short time periods and if they not correspond to the new positions obtained from the vision system, they can be recognized as moving objects, otherwise it can be assumed that objects are static and so we exclude further calculations on these objects. The idea is to implement heuristic algorithms in order to have quick decision, or even an exact algorithm in case the tests shown no impact by the heuristic algorithm. The algorithms should decide if navigation of robot should continue as planed or robot has to recalculate the path. These approaches distinguish two phases:

Learning Phase: Using of the multiple sensors, it is possible the obstacles and other static objects are recognized and visualized "on map". As a result there is one probabilistic road map when we differentiate between static and moving objects.

Decision Phase: Several heuristic algorithms are developed which calculates one more optimal path on already found map. Further, in this phase the data obtained from multiple sensors are compared them with the map objects in order to find an optimal moving decision towards local target. Decision Phase take in consideration the Learning Phase and records the "failed trying" in order to avoid the same path in case the previous path was not successful.

A very general schema of the phases of path planning approach is presented below. (Fig. 1.0).

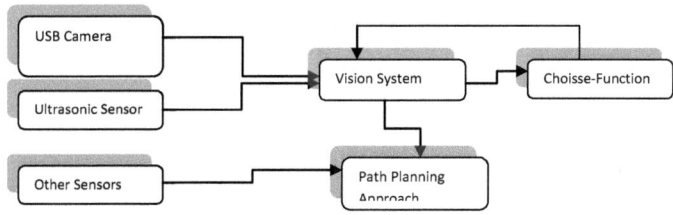

Fig. 1-0 General schema for different phases of path planning approach

Chapter outline

This thesis is divided into seven chapters. A brief description of the contents of each chapter is as follows:

Chapter 1 – Introduction into Autonomous Mobile robots and Humanoid robots .

Chapter 2 – State of the art of perception, localization, path planning and navigation of robots.

Chapter 3 – Presents of new cost oriented methods developed in this work.

Chapter 4 – Kinematic and Dynamic of "Archie" and "Humi" analysis.

Chapter 5 – Implementation and results Simulation of new developed approach in this work.

Chapter 6 – Future work.

Chapter 1

1 Introduction

The word Robot is introduced for the first time in year 1921 from Czech writer Karel Capek (Capek, 1920), where he denoted the robots in his play "R.U.R Rossum's universal robots" as automated workers that revolted.
The Robots can be divided into 4 main categories: manipulators, industrial robots, mobile robots and humanoid robots.

1.1 Autonomous Mobile Robots

The mobile robotics is a relative new (25 year old) field. The main development trends can be found especially in the last years, as it has been shown they can be useful in the different areas, like construction, agriculture and food industry, household, medical and rehabilitation application, industry or even for hobby (Kopacek, 2010).
Some basic characteristics of mobile robots are the mobility, its autonomy and perception ability. The robotics is a very complex and interesting field, which includes several mechatronics knowledge areas. In order to design an autonomous mobile robot, the following mechatronics knowledge is required:
- As a mobile robot needs locomotion in order to move, there is knowledge from mechanism, kinematics, dynamics and control engineering required. For example lot of research is done for mobile robots that can move, fly, jump etc.

- To understand the environment and to be able to percept it, the knowledge about sensors, computer vision and signal processing are necessary.
- Finally in order to navigate the robots should be able to locate and to follow the trajectories, what requires a large knowledge and research in computer algorithms, artificial intelligence, probability theory etc.

There are lots of examples where the mobile robots are irreplaceable. The mobile robot *sojourner* was used in the mission of exploring planet Mars in summer 1997 (**Fig. 1-1**). Although the robot was tele-operated from earth it was also equipped with external sensor that helped robot to detect obstacles.

Fig. 1-1 The mobile Robot sojourner used on the Mars mission in summer 1997
(Internet)

The Landmine Detection is another helpful using of mobile robots. There are different prototypes developed for this purpose. For example the IHRT institute (Vienna University of Technology) has developed a prototype of a demining robot (Silberbauer, 2008). This prototype consists of a platform and a metal detection sensor and is equipped with an internal micro controller as well as internal sonar sensors, position speed encoders and a battery pack for network-independent and autonomous operation. (Kopacek , 2010).

According to Kopacek (Kopacek , 2010), *a commercially available mine detecting set – produced in Austria - is attached on the robot basic-platform. This device is intended to detect land mines with a very small metal content (1.5g) 10cm below the surface of the ground and in fresh or salt water. The overall weight of the mounted sensor components is about 2.5kg.* (Kopacek , 2010) . The prototype of this six-wheeled robot named HUMI (Robot for humanitarian demining) is shown in (Fig. 1-2)

Fig. 1-2 HUMI **(Kopacek , 2010)**

Very interesting mobile robots as well as humanoid robots are developed also for hobby. Typical examples are robots that can play football. Although they are primarily constructed for sport and fun, the huge scientific work backs the developing of this technology, starting from perception and image processing through to applying artificial intelligence and propagation algorithms.

"For this purpose in June 1997 was founded also the Federation of International Robot-Soccer Association FIRA, with the basic goal of taking the spirit of science and technology of robotics to the laymen and the younger generation, through the game of robot soccer". (FIRA).

One of the reasons why this art of mobile robots, for example MiroSot Robots (Fig. 1-3) are very interesting, is among others its vision based multi agent system and also the relative fast moving in comparison to other mobile platforms (up to 3 m/s).

These 7,5cm x 7,5cm x 7,5cm robots fulfills their tasks as exactly as possible using the onboard controller, analog CCD or digital camera as vision system, radio- transmission and strategy module implemented in different algorithms (Putz, 2004) & (Kopacek, P. , Wuerzl, M. , Schierer, E., 2005).

Fig. 1-3 Mobile Mini Robot

1.2 Humanoid Robots

Humanoid robotics is one of the most exciting field and for lot of engineers and scientists it represents the most interesting task at all in robotics.

According to Kopacek (Kopacek , 2011).*A humanoid robot is a robot with its overall appearance based on that of the human body. Perception, processing and action are embodied in a recognizably anthropomorphic form in order to emulate some subset of the physical, cognitive and social dimensions of the human body and experience. Humanoid Robotics is not an attempt to recreate humans. In general humanoid robots have a torso with a head, two arms and two legs, although some forms of humanoid robots may model only part of the body, for example, from the waist up.*

Some humanoid robots may also have a 'face', with 'eyes' and 'mouth'. The definition of a humanoid is as simple as "having human characteristics." (Kopacek , 2011).

For many researchers the goal is to develop the humanoid robots capable to interact socially with humans and to help them in their everyday tasks. This art of robots should act and look like humans, should understand them, communicate, cook, cleans and why not, should be human's best friend. To create such extreme complex biped machine the large numbers of disciplines are affected and collaborate. Mechanical, electrical and computer science engineers, biologist, cognitive scientist, linguist, artists and lot of other scientist merge their knowledge and research for this purpose.

One of the first humanoid robots is *Elektro*, introduced on the world show New York in 1939. He was about 213 cm tall, 120 kg weight, could walk by voice command, speak about 700 words, smoke cigarettes, blow up balloons and move its head and arms. (Internet)

At (Kopacek , 2011) can be found that humanoid robots generally can be assigned into *professional humanoid robots*, *research humanoid robots* and *Humanoid "Toy" robots* (Kopacek , 2011).

- Professional humanoid robots are developed from big companies, with the idea of assisting humans on their task of everyday life. This art of humanoid robots is very expensive and only partially available on the market. Typical example QRIO (Fig. 1-4). (Kopacek, 2011)

Fig. 1-4 Humanoid Robot – QRIO

QRIO is a biped humanoid robot that is able to walk on uneven and sloppy surfaces, run, jump, perceive depth through its two CCD cameras, be able to create a 3D map of its surroundings, recognize people from their faces and voices, learn, connect to the internet via a wireless home network, download and read information in which you are interested about, sing, dance and survive a fall unscathed and get back up by itself again. (Kopacek, 2011)

- *Research humanoid robots* are usually prototypes developed from institutes and scientists in order to tests new ideas from mechanical engineering, computer science through to Artificial Intelligence. In (Fig. 1-5) is shown such humanoid robot developed and implemented at the Institute of Handling Robotics and Technology (IHRT) of Vienna University of Technology. This Humanoid Robot named Archie has a height of 150 cm, weight of less than 20 kg, with 2.5 h operating time by a walking speed 0.5 m / s and with 31 degrees of freedom. (Kopacek, 2011).

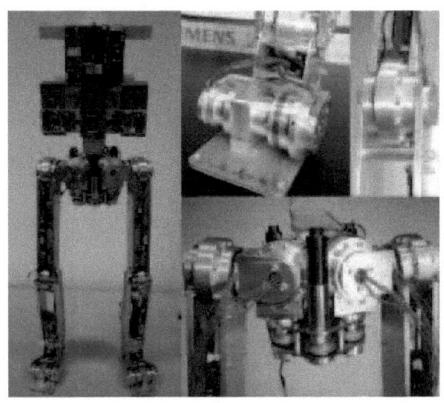

Fig. 1-5 Humanoid Robot - Archie

- *Humanoid "Toy" Robots* are the art of humanoid robots with limited capabilities from hardware and software point of view. The software installed on these robots usually is accessible and programming- able also in case there is not available higher level of programming knowledge, and can include graphical software interface with lot of preprogrammed behaviors. Typically they are equipped with ultrasonic distance sensors, light and sound sensors, ultrasonic range finder, touch sensors etc, and in some cases also with digital cameras. The can walk, climbs steps, does back flips, dance and also speech recognizing is available. (Kopacek , 2011).

An example for a reasonable cheap toy robot is the humanoid robot of Robonova (

Fig. **1-6**). It is a fully customizable and programmable aluminum robot. 16 digital servos and joints give complete control of torque, speed and position. It can walk, run, flips, cartwheels

Fig. 1-6 Humanoid Toy Robot – Robonova

and dance. (Kopacek, 2011). According to Kopacek (Kopacek, 2011), *The robot is available as a kit – assembly time approximately 8 hours - or as preassembled, ready to walk robot. In addition to walking, he can sense a wall using ultrasound. Robonova can be instructed to do cartwheels, take a bow and even do one-handed push-ups.*
(Kopacek, 2011).

Actually the main research areas in humanoid robotics are:

- Computer Vision and the variety of sensors, sensors fusion and perception which make possible for humanoid robot to percept and to track visually areas while moving.
- Interaction between humans and robots, especially the possibility of the humanoid robots to understand the gesture and facial expression of the humans.
- Machine autonomous learning techniques in order to enable robots to complete the given tasks unsupervised.
- Legged locomotion. Here is the stability one of the main tasks. The goal is to enable the humanoid robots to walk and to have the dynamic behavior similar to humans.
- Arm control and manipulation.

The main focus of this work is path planning and navigation. Therefore a brief introduction on robots path recognizing and path planning is given.

1.3 Navigation and Path planning

One of the main problems in mobile robotics and also in humanoid robotics field is the possibility of the robots to move autonomous in the environment, in order to fulfill tasks and to achieve in optimal way the goal position.

According to Ito (Ito,2009) *Mobile robots that perform their tasks in a real-world environment have one common problem to deal with: navigation. Navigation is a process which includes environment detection, path planning as well as path following. On the one hand, for local navigation the mobile robots have immediately to react to events indicated by sensors. This is necessary, for example, to drive through narrow passages and, in general, to avoid collisions with fixed and moving obstacles. On the other hand, for global navigation the mobile robots need some kind of map in order to reach a goal, which is too far away to be detected initially by the sensors.* (Ito, 2009).

In the literature, navigation can be classified either as indoor or outdoor navigation. The techniques based on map environment are used in indoor navigation and can be classify as follows: (Ito, 2009).

- *Navigation with maps.* Topological models of the environment created by the users are used.
- *Navigation based on map creation.* The fusion of sensors to generate the environment map that can be used to plan the robot's movements.
- *Navigation without map.* These systems do not use any environment representation but merely recognize objects that can be tracked.

For successfully navigating, robot has to interpret correctly the sensors data, i.e. to understand its *perception*. Further he should be able localize itself within the environment, i.e. to have a correct answer on the question *"where am I ?"*;. Furthermore he has to decide about the right actions to reach the goal position, i.e. to fulfill *cognition* problem. Finally the *motion control* should control drives to follow the planed trajectory.

When the mobile autonomous robot moves in a complex and difficult environment with no pre-knowledge of the area the finding of the optimal path is more difficult and time consuming. One of the best known methods of map building in an unknown environment is the SLAM method (Simultaneous Localization and Mapping). This method attempts to build a map of an unknown environment while simultaneously using this map to estimate the absolute position of the robot.

There is lot of research done for calculation of the optimal path (trajectory) using high sophisticated sensors and devices like GPS (Global Positioning System). They are still expensive and usually slow. The research mostly consists in finding or improving existing vision systems above all in image recognizing, as well as in creating of geometric path planner (GPP).

Because the sensors get faster and cheaper the robots can make decisions in nearly real time.

Different approaches are proposed for solving the path planning and navigation problem in two dimensions:

- Low dimensional problem: Different sophisticated algorithms like A^*, D^* or even heuristics and genetic algorithms can be used.
- For improving the performance, a so called coarse-fine approach can be applied (Kensuke, H. et al, 2010). These approaches compute first compute a coarse global path-plan continuously updated by the robot during the execution. Using additional information obtained from different sensors, velocity and obstacles for short horizons can be achieved a better performance.

In (Fig. 1-7) the coarse-planning provides a rough path, and the fine-planning is performed by a set of trajectories for vehicle dynamics.

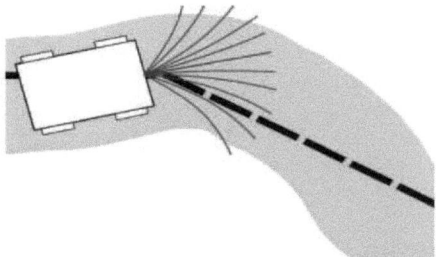

Fig. 1-7 Coarse-fine planning for a mobile robot **(Kensuke et al., 2010)**

The planning of the trajectory has high importance for robot navigation. Are the map given and the goal position known, the trajectory planning of the trajectory oft should be in function of the optimal path finding between start and the goal position.

The main requirements during the trajectory planning are the avoidance of the obstacles occurred on the trajectory, what can be denoted as *planning and reacting* problem. The planning and reacting approaches are complementary, i.e. the reacting without planning cannot guide the robot to reach the goal, as well as the planning without reacting is actually not useful.

The first step for path planning is the transforming of the continuous form of environment into a discrete map model. In general there are three different techniques:
- Road map, i.e. identification of a set of routes within the map.
- Cell decomposition, i.e. determining and fixing between free and occupied cells.

- Potential field, i.e. the robot is similar to the point under the influence of a potential field created from applying of different mathematical function over the space.

Chapter 2

2 State of the art

In this chapter the fundamentals for robust mobility are shortly outlined. Further an introduction, beginning from Perception, Localization, Path-Planning through to Motion control (Fig. 2-1) will be given.

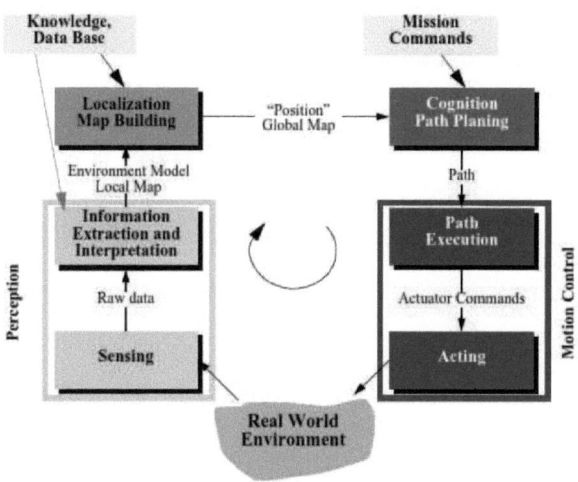

Fig. 2-1 Navigation of mobile robotics **(Siegwart,R. and Nourbakhsh.I.R., 2004)**

Patnaik wrote at (Patnaik, 2007).*Although the research in robot motion planning dated from late 60's during the early-stages of the development of computer controlled robots, most of the efforts is more recent and has been conducted during the 80's. In 1986, Rodney A. Brooks was the first to use the findings of ethological research, and to design a mobile robot. He published a seminar paper on the subsumption architecture, which was fundamentally a different approach in the development of mobile robots. He developed the subsumption language that would allow him to model something analogous to animal behaviors in tight sense-act loops using asynchronous finite-state machines. The first type of behavior for a robot was used to avoid obstacles that are too close and moving a little away or else standing still. Secondly, higher level behavior might be to move the robot in a given direction. This behavior would dominate the obstacle avoidance behavior by suppressing its output to the actuators unless an object comes too close. The higher levels subsumed the lower levels, and therefore the name of the architecture was subsumption architecture. They were able to develop a robot, using simple sonar or infrared sensors that could wave wander around a laboratory for hours without collisions with fixed objects or moving people. After this, Brooks and his colleagues developed highly intelligent mobile robots, i.e. robots, both wheeled and legged, which could chase moving objects or people and run or hide from light. Further, they can negotiate a cluttered landscape which might be found in a rugged outdoor environment.* (Patnaik, 2007).

Although in the last decade researcher develops more and more intelligent machines, still the main open problem in this field is perception. Perception is the main processes because the information obtained from the sensors and also sensor noise, impacts in other processes like localization, map-planning as well as motion control.

2.1 Perception

When the mobile autonomous robots move in complex open environments with no pre-knowledge one of the main task of this robot is acquiring of knowledge on this open environment. The only way to do this is through gathering of information using various sensors and then representation of this information as something "understandable" for mobile robot.

In this section the most common sensors as well as low cost sensors used in mobile and biped robots as well the extracting information from the sensors are introduced. The sensors are the inputs for the Perception process.

2.1.1 Sensors in Robotics (Siegwart and Nourbakhsh, 2004).

According to the literature the sensors which are responsible for measuring of the internal values of the robot such as motor speed or voltage etc are known as *internal* sensors. There are also *external* (*Exteroceptive)* sensors responsible for the acquiring of information from the environment such as distance measurements. There are also *passive* sensors like microphones or CCD cameras etc and the sensors who can achieve the best performance due the fact that they can emit the energy outside into environment and can measure the reaction known as *Active* sensors. Typical examples of active sensors are ultrasonic sensors, laser rangefinders. Some useful sensors used in robotics (Siegwart and Nourbakhsh, 2004) are shortly described.

2.1.1.1 Tactile sensors

These sensors are responsible for detection of physical contacts. Tactile sensors are very useful e.g. when the vision system cannot detect the obstacles. There are several types of tactile sensors, commonly force/torque sensors and thermal sensors.

2.1.1.2 Wheel motor sensors

These sensors are responsible for motor speed and position – Examples one: Brush encoders, Potentiometers, Optical encoders, Magnetic encoders etc. They can measure the internal state of a mobile robot.

2.1.1.3 Motion/speed sensors

The motion sensors can detect object and measure its speed when an object moves relatively to the robot. Typical examples are microwave and laser radar.

2.1.1.4 Vision-based sensors

For the humans the environment as such has a meaning because of the perception provided through human vision system. Through the human vision system the humans collects most of information about environment. For sensor technology is very important to create the sensors which are able to create and reproduce the capabilities of the human vision system. There are two most interesting technologies that are able to create vision sensors: CCD and CMOS. The first digital cameras used CCD to turn images from analog light signals into digital pixels. CMOS chips uses transistors at each pixel to move the charge through traditional wires. CCD sensors create the high quality images with low noise and are more sensitive to light.

2.1.1.5 CCD technology

Is the most popular basic ingredient of robotic vision systems today. The CCD chip (

Fig. 2-2) is an array of light-sensitive picture elements, or pixels.

Each pixel is a light-sensitive, discharging capacitor. As photons of light strike each pixel, they liberate electrons, which are captured by electric fields and retained at the pixel. Over time, each pixel accumulates a varying level of charge based on the total number of photons that have struck it. After the integration period is complete, the relative charges of all pixels need to be frozen and read. In a CCD, the reading process is performed at one corner of the CCD chip. The bottom row of pixel charges is transported to this corner and read, then the rows above shift down and the process is repeated.

Fig. 2-2 Cheap CCD camera (http://www.videologyinc.com/cameras/...)

The key disadvantages of CCD cameras are primarily in the areas of inconstancy and dynamic range. A number of parameters can change the brightness and colors with which a camera creates its image. Manipulating these parameters in a way to provide consistency over time and over environments, for example, ensuring that a green shirt always looks green, and something dark gray is always dark gray, remains an open problem in the vision community.

The second class of disadvantages relates to the behavior of a CCD chip in environments with extreme illumination. In cases of very low illumination, each pixel will receive only a small number of photons.

2.1.1.6 CMOS technology.

The complementary metal oxide semiconductor chip is a significant departure from the CCD. It too has an array of pixels, but located alongside each pixel are several transistors specific to that pixel. Just as in CCD chips, all of the pixels accumulate charge during the integration period CMOS has a number of advantages over CCD technologies. First and foremost, there is no need for the specialized clock drivers and circuitry required in the CCD to transfer each pixel's charge down all of the array columns and across all of its rows. This also means that specialized semiconductor manufacturing processes are not required to create CMOS chips. Therefore, the same production lines that create microchips can create inexpensive CMOS chips as well (Fig. 2-3). The CMOS chip is so much simpler that it consumes significantly less power; incredibly, it operates with a power consumption that is 1/100 of the power consumption of a CCD chip. In a mobile robot, power is a scarce resource and therefore this is an important advantage.

As compared to the human eye, these chips all have far poorer adaptation, cross-sensitivity, and dynamic range. As a result, vision sensors today continue to be fragile.

Fig. 2-3 Low-cost CMOS camera

2.1.1.7 Visual ranging sensors

Range sensing is extremely important in mobile robotics as it is a basic input for successful obstacle avoidance. As we have seen earlier in this chapter, a number of sensors are popular in robotics explicitly for their ability to recover depth estimates: ultrasonic, laser rangefinder, optical rangefinder, and so on. It is natural to attempt to implement ranging functionality using vision chips as well.

However, a fundamental problem with visual images makes rangefinding relatively difficult. Any vision chip collapses the 3D world into a 2D image plane, thereby losing depth information. If one can make strong assumptions regarding the size of objects in the world, or their particular color and reflectance, then one can directly interpret the appearance of the 2D image to recover depth. But such assumptions are rarely possible in real-world.

2.1.1.8 Stereo vision

Stereo vision is one of several techniques in which we recover depth information from two images that depict the scene from different perspectives.

2.1.1.9 Color-tracking sensors

Although depth from stereo will doubtless prove to be a popular application of vision-based methods to mobile robotics, it mimics the functionality of existing sensors, including ultrasonic, laser, and optical rangefinders. An important aspect of vision-based sensing is that the vision chip can provide sensing modalities and cues that no other mobile robot sensor provides. One such novel sensing modality is detecting and tracking color in the environment. For example, the annual robot soccer events make extensive use of color both for Environmental marking and for robot localization (Fig. 2-4).

Fig. 2-4 Color markers on the robot and color-tracking sensor

2.1.1.10 CMUcam robotic vision sensor.

This sensor is designed to provide high-level information extracted from the camera image to an external processor that may, for example, control a mobile robot. An external processor conFig.s the sensor's streaming data mode, for instance, specifying tracking mode for a bounded or value set. Then, the vision sensor processes the data in real time and outputs high-level information to the external consumer.

2.1.1.11 Active ranging sensors

These sensors continue to be the most popular sensors in mobile robotics. Many ranging sensors have a low price point, and, most importantly, all ranging sensors provide easily interpreted outputs: direct measurements of distance from the robot to objects in its vicinity. For obstacle detection and avoidance, most mobile robots usually use ranging sensors. But the local free-space information provided by ranging sensors can also be accumulated into representations beyond the robot's current local reference frame. Thus active ranging sensors are also commonly found as part of the localization and environmental modeling processes of mobile robots. It is only with the slow advent of successful visual interpretation competence that we can expect the class of active ranging sensors to gradually lose their primacy as the sensor class of choice among mobile robotic.

2.1.1.12 The ultrasonic sensor (time-of-flight, sound)

The basic principle of an ultrasonic sensor is to transmit a packet of (ultrasonic) pressure waves and to measure the time it takes for this wave packet to reflect and return to the receiver. The distance of the object causing the reflection can be calculated based on the propagation speed of sound and the time of flight .

The ultrasonic wave typically has a frequency between 40 and 180 kHz and is usually generated by a piezo or electrostatic transducer. Frequency can be used to select a useful range when choosing the appropriate ultrasonic sensor for a mobile robot. Lower frequencies correspond to a longer range, but with the disadvantage of longer post-transmission ringing and, therefore, the need for longer blanking intervals.

Ultrasonic sensors suffer from several additional drawbacks, namely in the areas of error, bandwidth, and cross-sensitivity. A limitation of ultrasonic ranging relates to bandwidth. For example, measuring the distance to an object that is 3 m away will take such a sensor 20 ms, limited its operating speed to 50 Hz. But if the robot has a ring of twenty ultrasonic sensors, each firing sequentially and measuring to minimize interference between the sensors, then the ring's cycle time becomes 0.4 seconds and the overall update frequency of any one sensor is just 2.5 Hz. For a robot conducting moderate speed motion while avoiding obstacles using ultrasonic, this update rate can have a measurable impact on the maximum speed possible while still sensing and avoiding obstacles safely.

2.1.1.13 Laser rangefinder (time-of-flight, electromagnetic)

The laser rangefinder is a time-offlight sensor that achieves significant improvements over the ultrasonic range sensor owing to the use of laser light instead of sound. This type of sensor consists of a transmitter which illuminates a target with a collimated beam (e.g., laser), and a receiver capable of detecting the component of light which is essentially coaxial with the transmitted beam. Often referred to as optical radar or *lidar* (light detection and ranging), these devices produce a range estimate based on the time needed for the light to reach the target and return. A mechanical mechanism with a mirror sweeps the light beam to cover the required scene in a plane or even in three dimensions, using a rotating, nodding mirror.

2.1.1.14 Ground based beacons

One elegant approach to solving the localization problem in mobile robotics is to use active or passive beacons. Using the interaction of on-board sensors and the environmental beacons, the robot can identify its position precisely. Although the general intuition is identical to that of early human navigation beacons, such as stars, mountains, and lighthouses, modern technology has enabled sensors to localize an outdoor robot with accuracies of better than 5 cm within areas that are kilometers in size.

2.1.1.15 The global positioning system (GPS)

Was initially developed for military use but is now freely available for civilian navigation. There are at least twenty-four operational GPS satellites at all times. The satellites orbit every 12 hours at a height of 20.190 miles are located in each of six planes inclined 55 degrees with respect to the plane of the earth's equator (
Fig. 2-5).

Each satellite continuously transmits data that indicate its location and the current time. Therefore, GPS receivers are completely passive but exteroceptive sensors. The GPS satellites synchronize their transmissions so that their signals are sent at the same time. When a GPS receiver reads the transmission of two or more satellites, the arrival time differences inform the receiver as to its relative distance to each satellite. By combining information regarding the arrival time and instantaneous location of four satellites, the receiver can infer its own position. In theory, such triangulation requires only three data points. However, timing is extremely critical in the GPS application because the time intervals being measured are in nanoseconds. It is, of course, mandatory that the satellites be well synchronized..

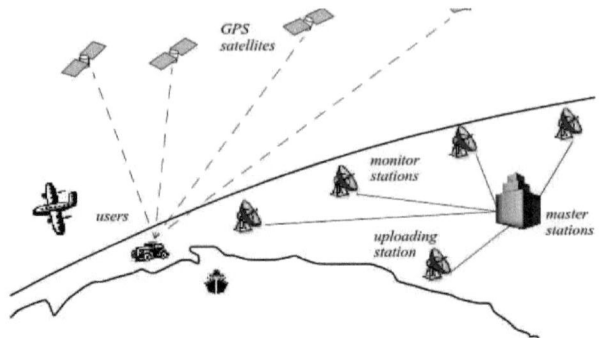

Fig. 2-5 Calculation of positions based on GPS.

2.1.2 Representing of errors and Uncertainty (Siegwart and Nourbakhsh, 2004)

To build a model of the environment, it is necessary the combining of the sensors, even of the same sensors but repeatedly over the time. There should be able to detect and measure the possible errors from single sensors, as well as the uncertainty of a whole robot system.

The error can be defined as the difference between sensor measurement and the true value. Let suppose the sensor has taken a set of n measurements with values ρ_i .

The goal is to characterize the estimate of the true value given these measurements:

$$E[X] = g\,(\rho_1,\rho_2,\ldots,\rho_n) \qquad (2.1)$$

The true value is represented by a random and unknown variable X. In (Fig. 2-6). can be found the *probability density function f(x)* for each possible value x of X.

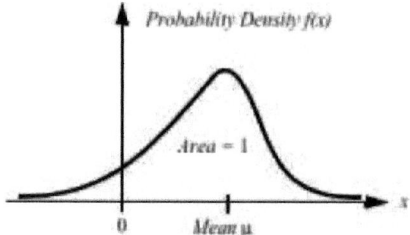

Fig. 2-6 Probability density function **(Siegwart and Nourbakhsh, 2004)**

The complete chance of X having some value is under the curve for Area=1,

$$\int_{-\infty}^{\infty} f(x)\, dx = 1 \qquad (2.2)$$

The probability of the X to be between two values a and b, can be computed as the bounded integral.

$$P[a < X \leq b] = \int_{a}^{b} f(x)\, dx \qquad (2.3)$$

Further is given the defining of the *mean, variance* and *standard* deviation:

Mean value μ is equal to $E[X]$ in case there will be infinite measuring of X , and the average value will be taken.

$$\mu = E[X] = \int_{-\infty}^{\infty} x\, f(x)\, dx \qquad (2.4)$$

Mean square value is the weighted average of the squares of all values of x.

$$E[X^2] = \int_{-\infty}^{\infty} x^2 f(x)\, dx \qquad (2.5)$$

The variance σ^2 represents the width of possible values of X.

$$Var(X) = \sigma^2 = \int_{-\infty}^{\infty} (x - \mu)^2 f(x)\, dx \qquad (2.6)$$

Standard deviation is the square root of the variance σ^2.

2.1.2.1 Gaussian Distribution

It is called also *normal distribution*. The advantages of Gaussian Distribution to other probability functions are:. (Siegwart and Nourbaksh, 2004)

- Its symmetry around the mean value μ
- It is unimodal, with a single peak that reaches the maximum at value μ

Below can be found the formula for the Gaussian probability density function

$$f(x) = \frac{1}{\sigma\sqrt{2\pi}} \exp\left(-\frac{(x-\mu)^2}{2\sigma^2}\right) \qquad (2.7)$$

According to Siegwart and Nourbaksh, (Siegwart and Nourbaksh, 2004) *In mobile robotics, one often uses a series of measurements, all of them uncertain, to extract a single environmental measure. For example, a series of uncertain measurements of single points can be used to extract the position of a line (e.g., a hallway wall) in the environment.*
Consider the system in Fig. 2-7, with n inputs-signals, m outputs-signals, and with a known probability distribution. To find the probability distribution of the output Y_i if they depend from known function f_i upon the input signals, as first the first order Taylor expansion of f_i will be used. (Siegwart and Nourbaksh, 2004)

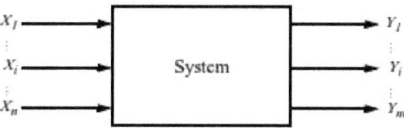

Fig. 2-7 Error propagation in system with n-inputs and m-outputs

The output covariance matrix C_Y is given by the error propagation (Siegwart and Nourbaksh, 2004)

$$C_Y = F_X C_X F_X^T \qquad (2.8)$$

where

C_X represents the input uncertainties

C_Y represents the propagated uncertainties for the outputs

F_X is the Jacobian Matrix

$$F_X = \nabla f = [\nabla_X \cdot f(X)^T]^T = \begin{bmatrix} f_1 \\ \vdots \\ f_m \end{bmatrix} \begin{bmatrix} \frac{\partial}{\partial X_1} & \cdots & \frac{\partial}{\partial X_n} \end{bmatrix} = \begin{bmatrix} \frac{\partial f_1}{\partial X_1} & \cdots & \frac{\partial f_1}{\partial X_n} \\ \vdots & \vdots & \vdots \\ \frac{\partial f_m}{\partial X_1} & \cdots & \frac{\partial f_m}{\partial X_n} \end{bmatrix} \qquad (2.9)$$

2.2 Localization

From all processes of robotics (see Fig. 2-1), the localization problem (Fig. 2-8) is one of the most researched area in past decades, and the most significant results are achieved here.According to (Siegwart and Nourbakhsh, 2004), In case of GPS the localization problem can be simplified. Although the GPS system can exactly determine the robot position, this method is not always appropriate. Due to the limitation, i.e. accuracy within to several meters, or what more the GPS system cannot function in indoors or in rough areas, this technology is not adequate especially for miniature mobile robots or body-navigating nanorobots

(Siegwart and Nourbakhsh, 2004). Furthermore not only the absolute position of the robot is important, but also the relative position to other objects is necessary for the robot, in order to achieve the goal position.

State of the art of the localization problem includes noise and aliasing problem, belief representations, map representation, probabilistic localization, as well as autonomous map building.

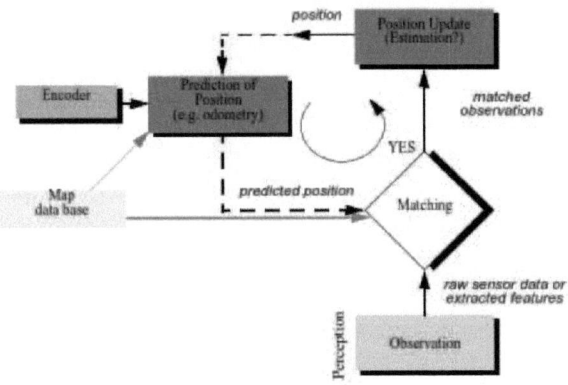

Fig. 2-8 General schema for Mobile Robot Localization
(Siegwart and Nourbakhsh, 2004).

2.2.1 Noise and Aliasing Problem (Siegwart and Nourbakhsh, 2004).

Sensors are the fundamental robot input for the process of *perception*, and therefore the degree to which sensors can discriminate the world state is critical. *Sensor noise* induces a limitation on the consistency of sensor readings in the same environmental state and, therefore, on the number of useful bits available from each sensor reading. (Siegwart and Nourbakhsh, 2004). In such cases can happen that environmental features cannot be captured and represented by the robots. A CCD camera for example can appear noisy in case of changing buildings illuminations. As a result, huge values are not constant. Picture jitter, blooming, blurring, signal gain etc can be additional source of noising affecting the content of video color image.

The sonar sensors for example when emits sound towards smooth surfaces most of the signals will be reflected way, so no echo can be generated.

The solution for sensor noise can be either multiple reading of sensors and/or fusion of sensors or multisensor data in order to increase the information quality for robot inputs. The main objective of sensor fusion is to collect measurements and observation from different sensors, to extract the information and to make the right decision

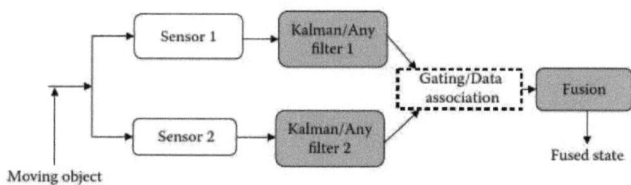

Fig. 2-9 Concept of sensor data fusion for target tracking **(Raol, 2010)**

2.2.2 Belief Representations (Siegwart and Nourbakhsh, 2004).

Supposing the robot has represented the model of its environment on the map. The first question is where is the robot localized on the map?, i.e. the robot must have also the representation of its belief regarding to the position on this map. In case there are more possible positions, the question is how these multiple are ranked.

At the single hypothesis be¹lief, the robots belief about its position on the map is expressed as single point on this map. In these cases there is no position ambiguity, as the robot assumed its position as correct, and all future actions are based on this assumption.

At multiple hypothesis belief, theoretically infinite possible positions can be assumed from robot as belief position. The possible representing of multiple hypothesis belief is it's mathematical distribution. The intended interpretation is that the distribution at each position represents the probability assigned to the robot being at that location. This representation is particularly amenable to mathematically defined tracking functions, such as the Kalman filter, that are designed to operate efficiently on Gaussian distributions. (Siegwart and Nourbakhsh, 2004).

The key advantage of the multiple-hypothesis representation is that the robot can explicitly maintain uncertainty regarding its position. If the robot only acquires partial information regarding position from its sensors and effectors, that information can conceptually be incorporated in an updated belief. A more advantage of this approach revolves around the robot's ability to explicitly measure its own degree of uncertainty regarding position. This advantage is the key to a class of localization and navigation solutions in which the robot not only reasons about reaching a particular goal but reasons about the future trajectory of its own belief state. (Siegwart and Nourbakhsh, 2004).

The main disadvantage of the multiple hypothesis belief is the making decision problem. If assumed the robots position is a set of possible position, the deciding approach, "what next"? is very hard problem, because there is a probability problem, and in general the right decision is computationally very expensive.

---

[1] Belief represents the expected position of the robot in the map, i.e. the position where the robot believes he is. The belief position in the real case corresponds not with the exact position.

2.2.3 Map Representation

The model of environment as map representation has an impact on the choices available for robot position representation. The precision of the map representation should match the precision needed from robot to achieve the goal position; the precision of information obtained from sensors as well as has an impact on computational complexity.

A very common method of environment representation is exact decomposition of the environment. To avoid the memory explosion only 2D mapping is done, where the obstacles are represented as polygons in continuous valued coordinate space. Representing of obstacles as such polygons will be done without storing any other features of objects, like color, texture etc, that have no direct impact on location of the obstacle on the map.

According to Siegwart in (Siegwart and Nourbakhsh, 2004).With radically decomposing environment into map, can be achieved the fidelity between real environment and the map. This can happen especially in case of an abstraction, i.e. the general decomposition and selection of environmental features. But in case the abstraction is planned carefully and decomposition is hierarchical, the map representation can be potentially minimized and the computationally can be far superior to planning in a fully detailed world model. (Siegwart and Nourbakhsh, 2004).

A very common decomposition strategy is *exact cell decomposition*. This method was introduced from (Latombe, 1991) where the decomposition is made by selecting boundaries between cells based on geometric extremities. There will be a drawing parallel line segments from each vertex of each interior polygon in the configuration space to the exterior boundary, where each cell is numbered and represented as a graphs node. (see Fig. 2-10) .

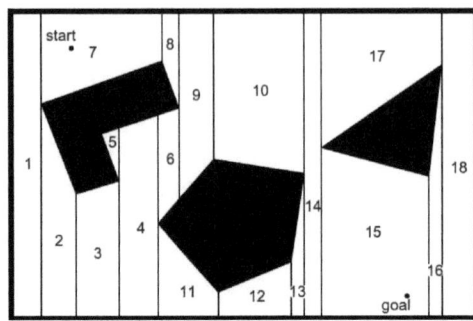

Fig. 2-10 Exact Cell Decomposition **(Siegwart and Nourbakhsh, 2004)**

Another decomposition approach is *fixed decomposition* where the continuous real environment will be transformed into discrete approximation. The disadvantage of this method is its inexact nature, i.e. it is possible to lose the passage-ways during the transformation.(see Fig. 2-11).

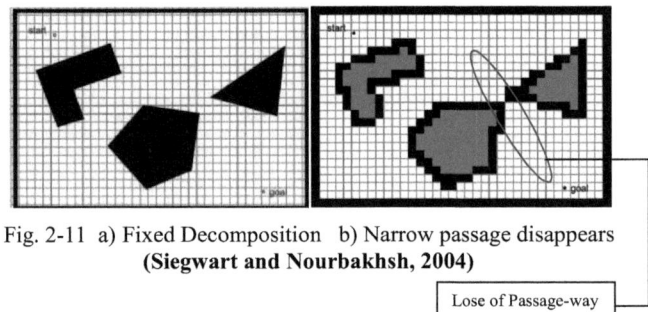

Fig. 2-11 a) Fixed Decomposition   b) Narrow passage disappears
(Siegwart and Nourbakhsh, 2004)

Most popular version of fixed decomposition is *occupancy grid*. Here the environment is represented as a grid of cells where each cell is either occupied or free. In first case the cell is part of an obstacle and can have a value 1 or true, in second case the cell is part of free space and can have value 0 or false.

There is also the *approximated cell decomposition* known also as "*quadtree*" decomposition. The cells will be recursively divided into four identical rectangles until one cell either is completely in free space or in configuration space of obstacle. In case that a predefined limit of resolution is achieved the recursion dividing will stop. (Fig. 2-12).

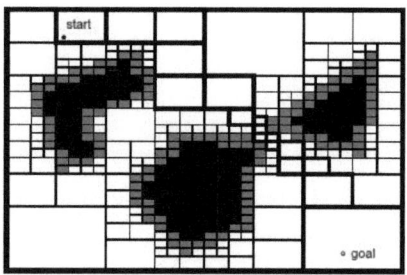

Fig. 2-12 "Quadtree" Cell decomposition **(Siegwart and Nourbakhsh, 2004)**

2.2.4 Probabilistic localization

The three major approaches are Markov Localization, Kalman Filter localization and Monte Carlo Localization. The complete introduction about these three approaches can be found a (Niemueller and Widyadharma, 2003), (Siegwart and Nourbakhsh, 2004) and (Ito, 2009). In this work only the short introduction is given.

At (Siegwart and Nourbakhsh, 2004) can be found the following: *These Localization methods explicitly identify probabilities with the possible robot positions, and for this reason these methods have been the focus of recent research. The first class, Markov localization, uses an explicitly specified probability distribution across all possible robot positions. The second method, Kalman filter localization, uses a Gaussian probability density representation of robot position and scan matching for localization. Unlike Markov localization, Kalman filter localization does not independently consider each possible pose in the robot's configuration space. Interestingly, the Kalman filter localization process results from the Markov localization axioms if the robot's position uncertainty is assumed* to have a Gaussian form. Kalman filter is an extremely efficient fusion. Assuming the data from two different sensors are taken, from first sensor at time k and from second sensor at time k+1. From each measurement there can be obtained estimated robots position, say q_1 and q_2. Using the Gaussian probability, it can be adopted follows:

$\hat{q}_1 = q_1$ with variance σ_1^2 and $\hat{q}_2 = q_2$ with variance σ_2^2.
Assuming there was no motion between time k and k+1, can be obtain the final form of Kalman implementation for fusion, in order to get the best estimate $\hat{q}$. (Siegwart and Nourbakhsh, 2004).

$$\hat{x}_{k+1} = \hat{x}_k + K_{k+1}(z_{k+1} - \hat{x}_k) \qquad (2.10)$$

where

$$K_{k+1} = \frac{\sigma_k^2}{\sigma_k^2 + \sigma_z^2}, \ \sigma_k^2 = \sigma_1^2, \sigma_z^2 = \sigma_2^2 \qquad (2.11)$$

The best estimate at $\hat{x}_{k+1}$ of the state x_{k+1} at time k+1 is equivalent to the best prediction of the value $\hat{x}_k$ before the new measurement z_{k+1} is taken, plus a correction term of an optimal weighting value times the difference between z_{k+1} and best prediction $\hat{x}_k$ at time k+1. The updated variance of the state $\hat{x}_{k+1}$ is given using equation (2.12) (Siegwart and Nourbakhsh, 2004).

$$\sigma_{k+1}^2 = \sigma_k^2 - K_{k+1}\sigma_k^2 \qquad (2.12)$$

A general introduction to Kalman Filter can be found at (Maybek).
There is also the Extended Kalman Filter (EKF) which is a continuous update cycle with alternating prediction phase and correction phase. The robots state X_t is calculated by its posterior probability with $P(X_t | z_{1:t}, a_{1:t-1})$ by a Gaussian. (Niemueller and Widyadharma, 2003).

The problem with Gaussian beliefs is its linear nature of motion model and its linear nature of measurement model. The linearization method used in EKF is Taylor expansion. In Fig. 2-13 is shown the localization of robot using EKF. The robot starts on the left, and then moved on dashed line. The ellipses show the uncertainty about its location, which is growing as result of Gaussian noise.

Markov localization tracks the robot's belief state using an arbitrary probability density function to represent the robot's position, see also (Burgard, Fox, Cremers, 1997) . According to (Siegwart and Nourbakhsh, 2004).*In practice, all known Markov localization systems implement this generic belief representation by first tessellating the robot configuration space into a finite, discrete number of possible robot poses in the map. In actual applications, the number of possible poses can range from several hundred to millions of positions.* (Siegwart and Nourbakhsh, 2004).

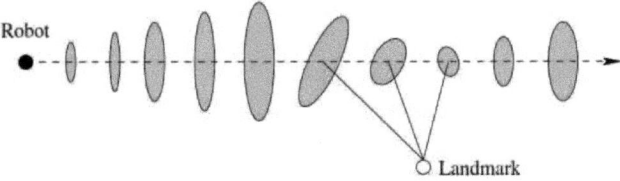

Fig. 2-13 Localization using the Extended Kalman Filter
(Niemueller and Widyadharma, 2003)

Finally, Monte Carlo Localization (MCL) is a particle filter algorithm. about the theory. The algorithm takes the given map and creates a population of N samples by a given probabilistic distribution. Then we start a continuous update cycle. The localization starts at time t = 0. Then the update cycle is repeated for each time step: (Niemueller and Widyadharma, 2003).

- *Each sample is propagated forward by sampling the next state value X_{t+1} given the current value X_t for the sample, and using the transition model given.*
- *Each sample is weighted by the likelihood it assigns to the new evidence, $P(e_{t+1}|x_{t+1})$*
- *The population is re-sampled to generate a new population of N samples. Each new sample is selected from the current population; the probability that a particular sample is selected is proportional to its weight. The new samples are un-weighted.*

After the enough knowledge about the environment is gathered after analyzing the information obtained from the range scanner, it resamples the population and particles concentrate at one or more point. So as the robot gathers more knowledge about the environment by analyzing the range scanner data it resamples the population and the particles concentrate at one or more points. At some point in time all points are in one cluster point. That is the location of the robot in the field. (Niemueller and Widyadharma, 2003).

2.2.5 Autonomous Map Building

One very interesting task in the robotics is the possibility of robots without help of humans to recognize its environment, to build the map, and to realize autonomously the given task. The complete introduction about this approach can be found at (Siegwart and Nourbakhsh, 2004).

"Accomplishing this goal robustly is probably years away, but an important sub-goal is the invention of techniques for autonomous creation and modification of an environmental map. Of course a mobile robot's

sensors have only a limited range, and so it must physically explore its environment to build such a map. The robot must not only create a map but it must do so while moving and localizing to explore the environment. In the robotics community, this is often called the simultaneous localization and mapping (SLAM) problem, arguably one of the most difficult problems specific to mobile robot systems. The reason that SLAM is difficult is born precisely from the interaction between the robot's position updates as it localizes and its mapping actions. If a mobile robot updates its position based on an observation of an imprecisely known feature, the resulting position estimate becomes correlated with the feature location estimate. Similarly, the map becomes correlated with the position estimate if an observation taken from an imprecisely known position is used to update or add a feature to the map. The general problem of map-building is thus an example of the chicken-and-egg problem. For localization the robot needs to know where the features are, whereas for map-building the robot needs to know where it is on the map". (Siegwart and Nourbakhsh, 2004).

2.3 Path Planning and Navigation

Assuming that the localization of robot on map is known, the path planning deals with the finding of the best possible path between start and target position within this map.

According to (Gonzales,2008*) this task is extremely difficult if no prior information is available and is trivial if perfect prior information is available and the position of the robot is precisely known. Perfect prior maps are rare, but good-quality, high-resolution prior maps are increasingly available. Although the position of the robot is usually known through the use of the Global Position System (GPS), there are many scenarios in which GPS is not available, or its reliability is compromised by different types of interference such as mountains, buildings, foliage or jamming. If GPS is not available, the position estimate of the robot depends on dead-reckoning alone, which drifts with time and can accrue very large errors. Most existing approaches to path planning and* *navigation for outdoor environments are unable to use prior maps if the position of the robot is not precisely known. Often these approaches end up performing the much harder task of navigating without prior information.* (Gonzales,2008)

When we speak about path trajectory planning it can be differentiate between several forms of trajectory planning. At (Gonzales,2008) can be found that Classical trajectory planning is dealing with the problem of finding the best path between two positions, assuming the position of the robot is known at all times. Because of the deterministic nature of the problem, and the fact that good heuristics can be easily found, the problem can be efficiently addressed using deterministic search techniques. In general the problem consists in modeling the state space as a graph $G(V,E)^2$, and then finding an appropriate search algorithm to search such graph (Gonzales , 2008)

Classical trajectory with uncertainty in position is another approach who is dealing with uncertainty in position, classical path planning algorithms usually model uncertainty as a region of uncertainty that changes shape as states propagate in the search. (Gonzales , 2008).

According to (Gonzales,2008) *One of the first planning approaches with uncertainty was called pre-image back chaining. This approach was designed from (Lozano-Perez, Mason and Taylor, 1983). Lazana and Latombe (Lazanas and Latombe, 1992), later expanded the pre-image back-chaining approach to robot navigation. For this purpose they define the bounds in the error of the position of the robot as a disk, and they use an uncertainty propagation model in which the error in the position of the model increases linearly with distance traveled. The idea of a landmark is also introduced as a circular region with perfect sensing. The world consists of free space with disks that can be either landmarks or obstacles* (Gonzales , 2008). Another approach where landmarks in the environment are used, and where landmark ability to localize the robot and Dijkstra

---

[2] G(V,E) is formal definition of the graph, where V denotes vertices and E denotes edges.

algorithms tries to find a path which minimize the uncertainty can be found at (Takeda, Latombe, 1992).
The first step in path planning is the possibility of transforming of continual environment into discrete map for any chosen path planning algorithm. The three main strategies for decomposition are: (Siegwart and Nourbakhsh, 2004).

- Road Map – where the set of routes within free space will be identified
- Cell Decomposition – Discriminate between free and occupied cells
- Potential Field – using of mathematical function over the space

Further, the different Road Map approaches are introduced. Cell Decomposition methods are already introduced in section 2.2.3.

2.3.1 Road Map Approaches (Siegwart and Nourbakhsh, 2004).

The basic idea of a road map approach is the possibility to connect the free space of configuration with one dimensional lines called as road maps. These road maps can then used as network of paths for robot motion planning from start to target position. Typical road map approaches are *Visibility graph* and *Voronoi Diagrams* . In the first case the paths should be as close as possible to obstacles, in second case the paths should be as far as possible from obstacles. The *Visibility Graph* is very easy for implementing, especially in case of polygon representation of obstacles. Although this method is optimal in terms of the length of the solution of paths, is not so optimal in case the density of obstacles is large. There is also safety problem, as the roads are too near to the obstacles. (Fig. 2-14).

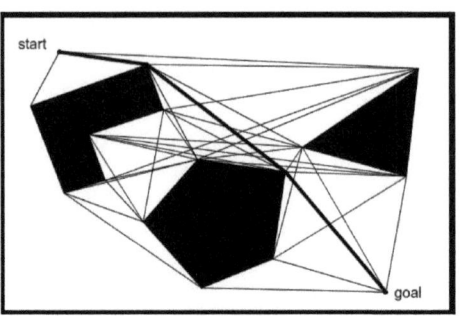

Fig. 2-14 Visibility Graph **(Siegwart and Nourbakhsh, 2004)**.

Voronoi Diagram is the opposite from *Visibilty Graph* , as the approach tends to maximize the distance between robot and obstacle. The connected cells tend to be equidistant from two or more obstacles (see Fig. 2-15). In case of polygon obstacles the Voronoi-paths are straights or parabolic segments. The weaknesses of this method are not optimality of length of path, as well as not possibility for short range sensors to percept the environment.

Probabilistic Roadmap Method . At this method a number of random points in free space are generated and each point is connected to nearest neighbor with paths like straight line overcoming any obstacle. As result, the approach tends to create a graph with no cycles. The main advantage of this approach is that only a low number of points should be tested to understand if the paths between these points are obstacle-free.

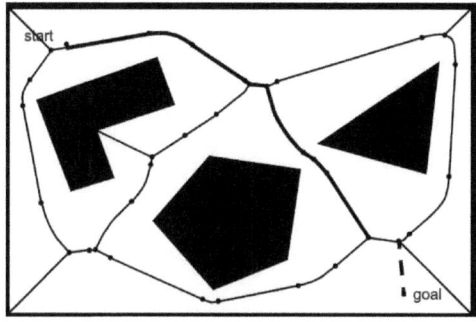

Fig. 2-15 Voronoi Diagram **(Siegwart and Nourbakhsh, 2004)**.

D* approach is a very popular algorithm used for real-world application. At this application the occupancy grid will be viewed also as a cost map, where cost c is equally to traversing cell horizontally and vertically. In case of obstacle the cost is ∞. The D^* minimize the total cost of the travel. The main advantage is the possibility of replanning of the path in case it is shown the cost is greater than expected, similar to backtracking.

2.4 Obstacle Avoidance – BUG Algorithm

The BUG Algorithm is the simplest intuitive obstacles avoiding Algorithm. There are two different but very similar approaches. At BUG1 the main idea of this algorithm is only to follow the whole contour the obstacle in order to find the shortest possible way to goal. After the whole contour is followed, the robot moves from the point where the shortest path to target is. At BUG2 the difference from BUG1 is that here, the algorithm stops with following of the contour immediately after the robot finds a possibility to move directly to goal (Fig. 2-16).

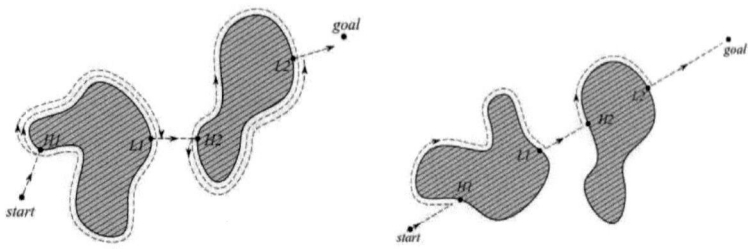

Fig. 2-16 a) BUG 1 Obstacle Avoidance Algorithm   b) BUG 2 Obstacle Avoidance Algorithm

(Siegwart and Nourbakhsh, 2004)

Chapter 3

3 A new cost oriented method

The main goal of this work is to present a new cost oriented method for:

- Perception and building of the map of the environment,
- Localization of the robot,
- Path planning by calculating of the free spaces and trajectory simultaneously.

The important task are approaches of the process of building of a qualitative map, especially the detecting of already explored objects and obstacles which remain in new iterations of map recalculation. The process of differentiating between fixed and moving obstacles, during map calculation and path planning is another task. For reactions of a mobile robot in front of obstacles different algorithms are developed. All these are derived modeled and the software described and implemented.

The main advantages of these methods are the reduced time of map-calculation and the replacement of expensive high-tech with cost-oriented sensors (COS).

The main parts of proposed new software concept are shown in Fig. 3-1 :

- Vision System as part of perception that is able to detect and recognize objects;
- Map-Calculating of environment
- Localization of mobile robot
- An approach of path planning and optimal robot navigation using heuristic and exact algorithms;

- Simulation

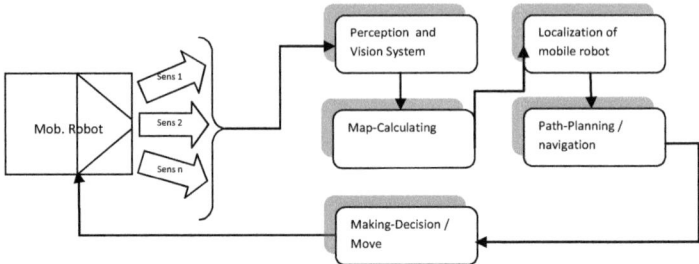

Fig. 3-1 Software development concept in general

These modules will be described more detailed.

3.1 Perception and Vision System

The perception is a process of two steps. The first step consists by the collecting of the data from the different sensors (visual sensors, i.e CCD camera), and the second step is converting of these measured data in something "mean-feeless" for the robot.

For the perception the ultrasonic sensors, light sensors and touch sensor are used as non-vision sensors as well as a CCD Camera as a vision based sensor. The perception is based on the data from the ultrasonic sensors data. The images obtained from a CCD camera serve as second range data and are in function of improving of perception.

The ultrasonic sensors are another type of sensors that can "see" the environment and objects. These sensors enable the robot to recognize objects, to measure distances, to detect movements and to avoid the obstacles. The distance ranges in this work are between 0 to 3 meters.

The light-sensors are used for the robot in order to distinguish between darkness and light in the working environment or to mess of the reflecting light-intensity from surfaces.

As another type of sensors that can be used in this work are touch sensors, which should avoid any direct contact between robot and object in case the ultrasonic sensor and CCD couldn't provide the exact information about position or size of the object.

The primary data are here from the ultrasonic sensors. All data from other sensors serve for improving or correcting.

Because different sensors get different data-information like range, size, angle, force etc, and because they can "offer" only the part of the environment, data-fusion is necessary.

Sensor fusion means the combination of the data coming from sensors such that the information obtaining after the fusion should be better than any information obtained from one individual sensor.

The sensor fusion made possible the failure-tolerance, i.e. in case some data from any sensor is going lose, it is possible through other sensor (redundant data), to recover or to improve the information. Another performance of sensor fusion is that some sensor information can be "confirmed" from other sensor information. Among lot of other improving, the main benefit of sensor fusion is that the information is continuously available. Fig. 3-2 shows general schema of the sensor fusion processes as well as the process of data-obtaining from different sensors proposed in this work.

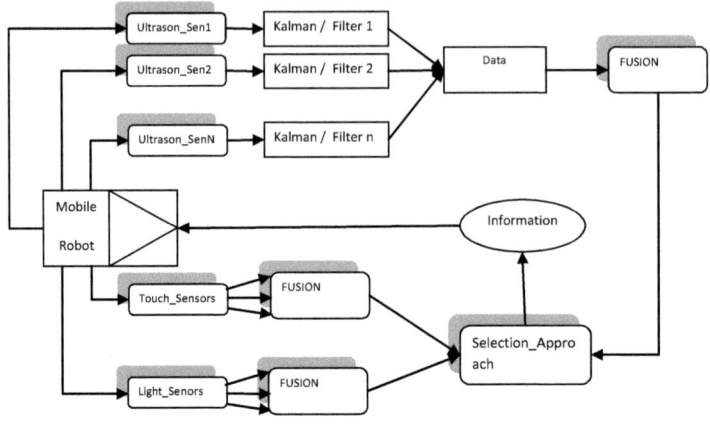

Fig. 3-2 Sensor Fusion System

There is a selective fusion process between the sensors of the same type, i.e. ultrasonic sensors, as well as another fusion for touch sensors and light sensors. There is no fusion between different kinds of sensors. Instead a selection-approach is used. This approach in respect to calculated evaluate-function corrects the certain information.

In this work a *Kalman-Filter* as a Fusion-algorithm is proposed. This algorithm in the literature is known as one of the best known approach for the fusion problems.

In (Gan, 2001) and (Huosheng, 2005) it is shown that in case the different sensors have identical measurements matrices, a better method for fusion is to combine the multi-sensor data based on minimum-mean-square-error. The computational load is lower, although the method where are used all raw measurement information seems to be exacter method.

The perception through CCD camera is another technique proposed in this work.

Fig. 3-3 show the general scheme about the procedure between camera image of object and its pixel presentation on the pixel coordinate system.

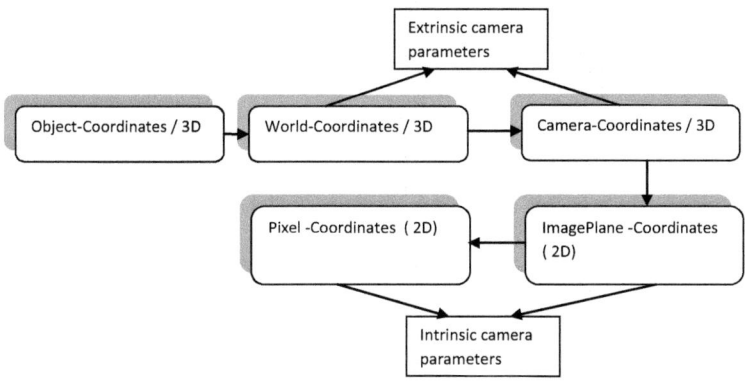

Fig. 3-3 Image projection from 3D into 2D pixel coordinates

The Extrinsic[3] camera parameters identify uniquely the transformation between the unknown camera reference frame and the known world reference frame.

The Intrinsic[4] camera parameters characterize the optical, geometric and digital characteristic of the cameras. In OpenCV and also Matlab offer the possibility these camera parameters to calibrate. In this work these functions are proposed to be used.

Another issue during the image processing is the question how to isolate the object from the rest of the image. Because in this work, of interest are the objects max. 3 meter way more than other objects in background, the method proposed in this work are the Background subtraction and the Color segmentation. To find the different homogenous regions on the image (blobs) in order to identify different objects, there is also blob-method proposed.

---

[3] Means the finding of translation vector that maps camera's origin to the world's origin as well as finding of the rotation matrix that aligns the camera's axes with world's axes

[4] Typically deals with the perspective projection, transformation between image plane coordinates and pixel coordinates as well as geometric distortion introduced by optics

Fig. 3-4 shows block diagram of the robot perception. Here, the sensors data are obtained after fusion processes and after the selection-approach is applied. These information and the images obtained from different camera-frames are then the basic information for the next step, the Map-Calculating.

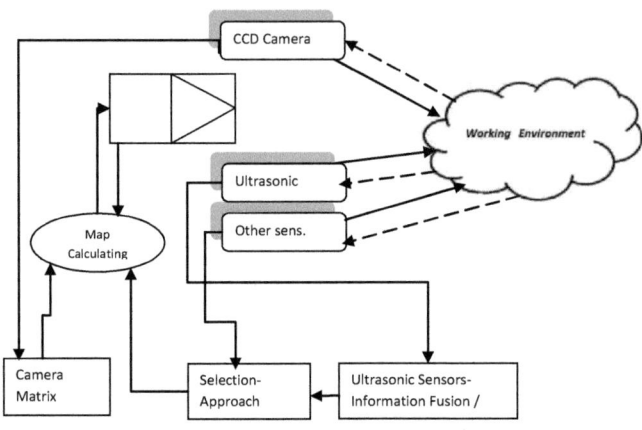

Fig. 3-4 Perception system in general

3.2 Map calculating of the environment

In previous section the COR vision perception was presented. The information's obtained from this process are the necessary basic inputs for the map-calculation. The terms like neighbor pixels or mapped-objects etc will be defined in order to implement the image-processing algorithms. Finally the pseudo-algorithms for the map-calculating will be shown.

The problem of mapping of environment and self localization of the robot are one of the most curiosity problems in mobile robotics. These two processes are also fundamental prerequisites for many other tasks like trajectory planning and navigation. The building of map has a direct impact on accuracy of the robot and on the calculations during the localization and navigation process of the robot. This is based on COR sensors.

Here an *exact decomposition* of the working environment is used. The concept of *fixed decomposition* and *exact decomposition* is widely used in mobile robotics and it's perhaps the most frequently used map representation technique currently. In order to reduce the calculations time a map-representation only in 2D (pixel coordinate system) is used. There are several possibilities and extensions of this form of representation. One possibility is to use a 2D representation in which polygons represent all obstacles in a continuous-valued coordinate space similar as used by Latombe (Siegwart and Nourbakhsh, 2004) . Another possibility is to use simple convex polygons without any another information, so it is achieved to reduce the memory usage (Siegwart and Nourbakhsh, 2004). Because we are not interested in complete information of our basic objects, an abstraction i.e. a version of *fixed decomposition* named *occupancy grid* where every binary cell is either filled or empty (Fig. 3-6) is presented. Here an adoption and extension of the *occupancy grid* with other data-information where it is exactly known for every filled cell the "status" of its neighbor cells. In this way we can detect the discontinuities of filled cells in order to create all separate obstacles. It seems to be something similar to *topological decomposition*, but in our case the nodes are the objects (obstacles) instead specific areas as they are in topological decomposition, and connectivity between two nodes (objects) denotes the shortest distance between two objects. All these data are memorized in the data-structure, and although at the beginning the usage of memory is higher, it is helpful in case of trajectory calculations and especially in case of detecting "moving" objects within working area.

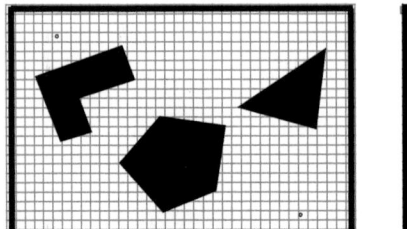 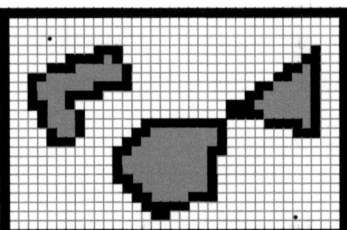

Fig. 3-5 Fixed Decomposition of the space   Fig. 3-6 Occupancy grid
(See section 2.2.3 and Fig. 2.11)

When the robot has no knowledge about the environment the first approach is the initializing of the maps (*pixel coordinate system*) by the defining of *image plane*, and *world-coordinate frame* (WCF). The map of environment we represent as a grid of cells (pixels) $cell_i = \{x_i, y_i, F_i\}$ where each cell has a triple of coordinates *x*, y and an information value *F*. According to this map, the image is represented as a grid of integer values from top-bottom and left-right.

F is a Boolean variable FALSE (0) or TRUE (1), and denotes if some obstacle or part of it is within that cell or not. In case of "TRUE" value the cell is filled, otherwise the cell is empty. Each cell (pixel) has integer coordinates (Fig. 3-7). The origin of world-coordinate frame is a position of mobile robot, so using the extrinsic and intrinsic camera parameters it's follows the transformation of WCF into image plane and finally into pixel coordinate system. For this purpose it is proposed openCV and (or) Matlab
.

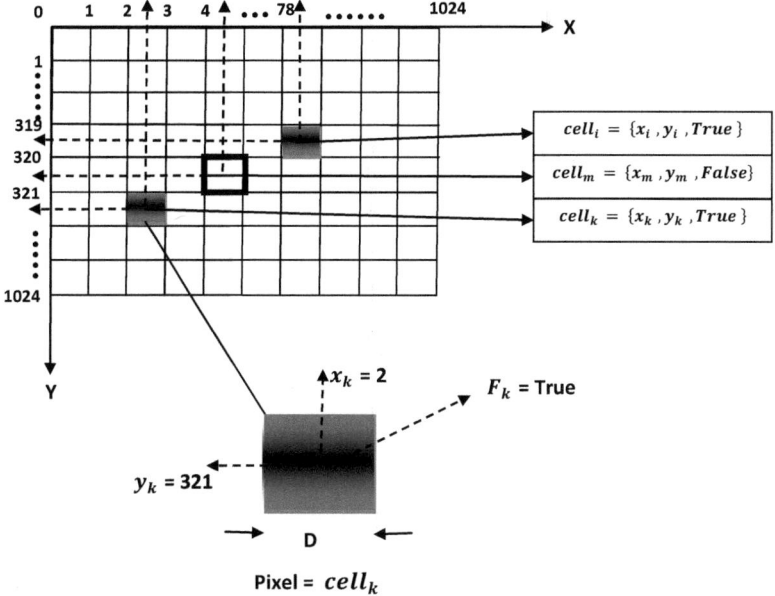

Fig. 3-7 Cell of the map-grid in pixel frame coordinate-system

In Fig. 3-7 it is shown a pixel-frame system 1024 x 1024 pixel[5], where for example $cell_i = \{x_i, y_i, F_i\}$ has coordinates $x_i = 78$, $y_i = 319$ and $F_i = True$, $cell_m = \{x_m, y_m, F_m\}$ has coordinates $x_m = 4$, $y_m = 320$ and $F_m = False$, $cell_k = \{x_k, y_k, F_k\}$ has coordinates $x_k = 2$, $y_k = 321$ and $F_k = True$.
D denotes the size of the cell, i.e the pixel dimension.

The relation between Cell-indexes and the Cell-coordinates is shown in Fig. 3-8 . If X_{max} is denoted as the max value of the X-axe (In Fig. 3-8 is 1024), then for any $cell_k$, the relation between index k and coordinates x_k and y_k is $k = X_{max} * y_k + x_k$.

---

[5] In general it depends from camera resolution. In this work, 1024x1024 pixel frame is taken only as example.

From here follows that $x_i = i \mod X_{max}$ *and* $y_i = i / X_{max}$.

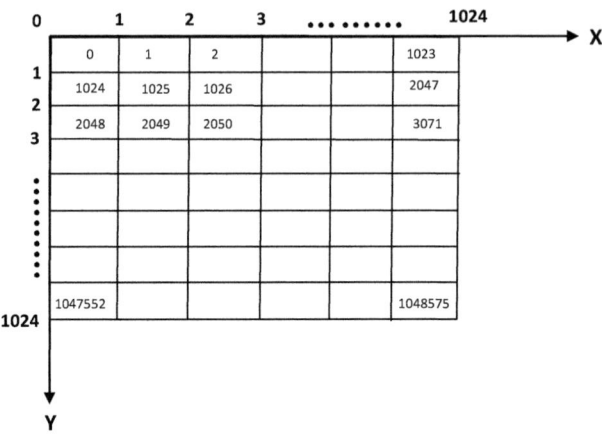

Fig. 3-8 Cell-indexes and the Cell-coordinates

Fig. 3-9 shows that for every cell it is possible to find its F- value (i.e. if it is filled or empty). The data are stored on an extra Matrix (Binary Image Matrix, Fig. 3-9) where the positions are one-to-one in correspondence with pixel-frame system.

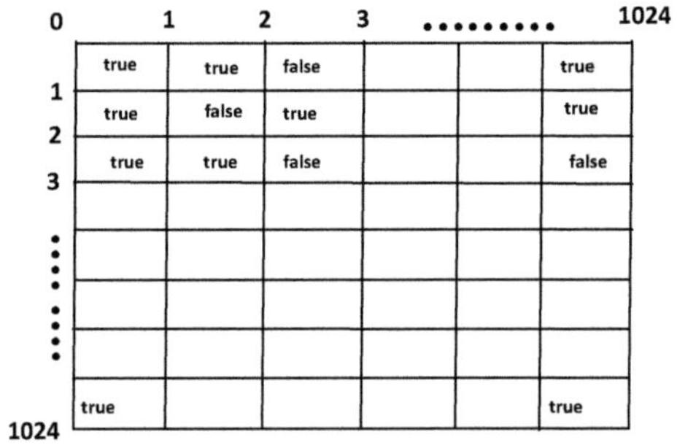

Fig. 3-9 Binary Image Matrix

Below can be found some basic definitions and some very general pseudo-codes in relation to these definitions:

Definition 3.1: Cell $cell_i = \{x_i, y_i, F_i\}$ is *neighbor* of another $cell_k = \{x_k, y_k, F_k\}$, if $|x_i - x_k| \in \{D_x, 0\}$ and $|y_i - y_k| \in \{D_y, 0\}$ where D_x and D_y are the calibration size for a map-pixel.
Formally: $\mathbb{N}(cell_i, cell_k)$ if $\{|x_i - x_k| \in \{D_x, 0\} \wedge |y_i - y_k| \in \{D_y, 0\}\}$ (see Fig. **3-10** a).
From definition follows that $\mathbb{N}(cell_i, cell_k) \equiv \mathbb{N}(cell_k, cell_i)$

In this definition the 8-neighbour model is used.

Definition 3.2: Cells $cell_i = \{x_i, y_i, F_i\}$ and $cell_k = \{x_k, y_k, F_k\}$, are *directly-connected* if they are neighbors and both are filled, i.e F_i and F_k are TRUE.
Formally, $\mathcal{D}_c(cell_i, cell_k)$ if $\{\mathbb{N}(cell_i, cell_k) \wedge F_i = F_k = TRUE\}$, (see Fig. 3-10 (b))

From definition 2 follows that $\mathcal{D}_c(cell_i, cell_k) \equiv \mathcal{D}_c(cell_k, cell_i)$

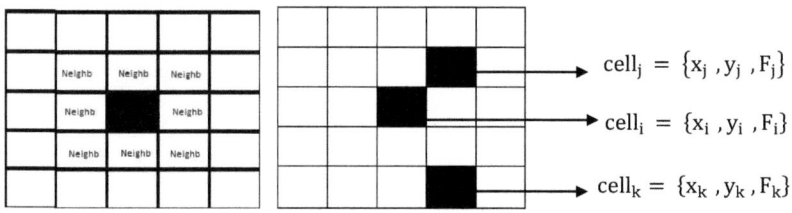

Fig. 3-10 a) Cell-Grid and all its neighbors   b) Directly-connected cells: .i.e **cell$_i$** and **cell$_j$** are directly connected, but no-one of them is directly connected with **cell$_k$**

The pseudo code for finding of all Direct - connected cells $\cup$ $cell_k$ of any certain cell $cell_i$ $\{x_i, y_i, F_i\}$, if $\mathcal{D}_c$ ($cell_i$, $cell_k$) is

```
//The Runtime of the Algorithm is linear, i.e O(n).
List Find_All_DirConnectCells ( Cell_i )
{
    1. Find all neighbors coordinates of coordinates x_i, y_i ;
    2. For each pair of coordinates x_new, y_new check the value in
       Binary Image Matrix;
    3. For each "true" value, store the coordinate into List;
    4. Return List;
}
```

Definition 3.3: Cells $cell_i = \{x_i, y_i, F_i\}$ and $cell_p = \{x_p, y_p, F_p\}$, are *continually-connected* if they are either *directly-connected* or there exists at least one cell, i.e $cell_j = \{x_j, y_j, F_j\}$ which is *continually-connected* with both of them.

Formally :

$\mathbb{C}_c$($cell_i$, $cell_p$)   if

$\{\mathcal{D}_c\,(\,cell_i, cell_p\,) \vee \exists\, cell_j \mid \mathbb{C}_c(\,cell_i, cell_j\,) \wedge \mathbb{C}_c(\,cell_p, cell_j\,)\,\}$

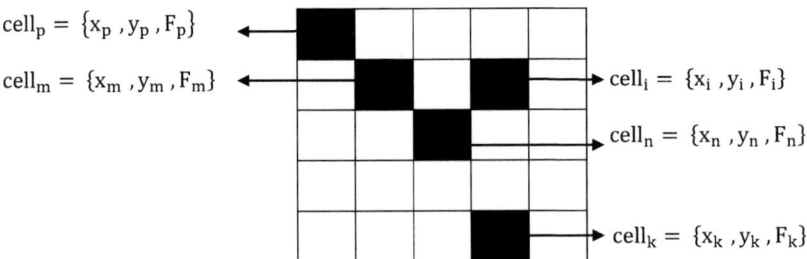

Fig. 3-11 $cell_i = \{x_i, y_i, F_i\}$ and $cell_p = \{x_p, y_p, F_p\}$ are continually-connected

From Fig. 3-11 it can be shown that $cell_i$ and $cell_p$ are continually connected, because as per definition, there exists $cell_m$ that is *directly-connected* with $cell_p$ and there exists also $cell_n$ *directly-connected* with $cell_m$ and $cell_i$.

The pseudo code (similar to Breadth-First Search Algorithm) for finding if two cells $cell_i$ and $cell_k$ are Continually-Connected is:

```
//The Runtime of the Algorithm is in case of adjacency matrix quadratic
//O(n²).
Boolean Are_CC (Cellᵢ , Cellₖ )
{
    1. Find all Direct-connected Cells of Cellᵢ , DirConnectedCells
    2. For each cell  Cellₘ  ∈ DirConnectedCells
    3. If Cellₘ = Cellᵢ , then Return True,
       ELSE
    4. Call Function recursively with new parameter  Return Are_CC
       (Cellₘ , Cellₖ )

}
```

Definition 3.4: mapping_object is denoted as the union of all *continually-connected* cells of map if every cell of *mapping_object* is *continually-connected* with all other cells of that mapping-object. (see Fig. 3-12)
Formally:

$$map\_object = \{ \bigcup_n^m cell_i , \forall\, cell_p, cell_s \in map\_object \Rightarrow \mathbb{C}_c(\,cell_p, cell_s\,)\}$$

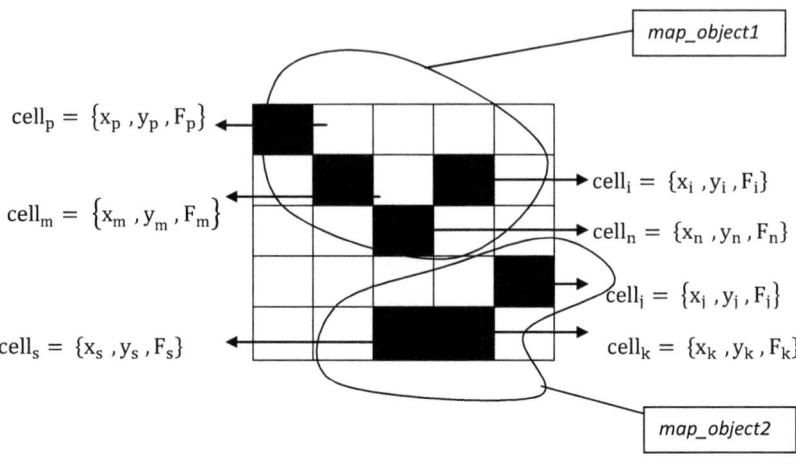

Fig. 3-12 Two different form of objects (map_object1 and map_object2) found

The pseudo-code for finding of the mapped-object is:

//The Algorithm for MapObject
IntegerList *Create_MapObject* (int x)
{
    1. *Find all Direct-connected Cells of* $Cell_i$, *DirConnectedCells*
    2. *Find the Union of Cells between the List DirConnectedCells and result list Result-List*
    3. *Update the Result-List with the results from step 2*
    4. *For each cell* $Cell_k$ ∈ *DirConnectedCells* , *call the function* **recursively** *for its index k.* => *Create_MapObject* (k)
    5. *Return Result-List*
}

The following pseudo-code algorithm finds all mapped-objects on the environment-map, beginning from any $cell_i$ where $F_i = True$

//The Algorithm for finding of all objects
ListOfObjects *FindAllObjects* ($cell_i$)
{

1. Create a first mapped-object, found_object = ObjectCreate_MapObject (i) and add it into ListofObjects.
2. For all ($cell_k$ member of *Binary Image Matrix* and true, i.e if F_k = True and not already checked)
3. In case $cell_k$ is not member of the found objects , i.e If ($cell_k$ ∉ ∪ found_objects)
4. Call the function **recursively** for its index k. Create_MapObject (k), and mark $cell_k$ as checked.
5. Return *ResList;*

}

Definition 3.5: Let be $cell_i$, $cell_j$ ∈ $mapping\_object$, than $cell_i$ and $cell_j$ are in *direct_relation*.

Formally: $\mathcal{DR}(cell_i, cell_j)$.

In case they are Not in *direct_relation*, than formally $\neg\mathcal{DR}(cell_i, cell_j)$.

Definition 3.6: Let be $cell_i$ ∈ $mapping\_object_x$, and let exist $cell_k$, so that $\mathbb{N}(cell_i, cell_k)$ and $\neg\mathcal{DR}(cell_i, cell_k)$. In that case, $cell_i$ is called *boundary_cell* of *mapping_object* .

Formally:

$\mathcal{BC}_i(mapping\_object_x) \equiv$

$cell_i$ if $\exists\ cell_k$, $\mathbb{N}(cell_i, cell_k)$ ∧ $\neg\mathcal{DR}(cell_i, cell_k)$, (see Fig. 3-13).

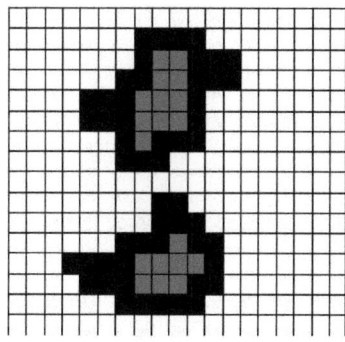

Fig. 3-13 Black colored pixels show the boundary cells (pixels) of two different objects

The following pseudo-code algorithm finds all boundary-cells of one mapped-object object_x

// *The Algorithm for finding of all boundary cells. The Runtime of the Algorithm is linear.*
IntegerList *Find_All_BoundaryCells (*Object object_x *)*
{

    1. *For each cell member of the object, i.e* $Cell_k \in$ object_x
    2. *Find all neighbor-cells of cell* $Cell_k$
    3. For each $Cell_n$ member of neighbor-cells
    4. Check in Binary Image Matrix If $F_n =$ False .
    5. If false then mark $Cell_k$ as boundary cell and break the loop.

}

Although the finding of the separate objects in this way leads to the unique detection of the objects, sometimes it can be not so useful separating of the image on very large number of the objects with the small size.

The first reason is the possibility that in fact these small-sized objects can be part of the homogenous object, but due to the errors cannot be presented as such. These errors can be occurred from the noise ore changing light intensity.

Another reason is the non-sense of presenting of such big number of objects although they are so "near" to each other. Why should be take in consideration lot of calculation, if between these objects due to the small distance no COR can move.

A homogenous object is:

Definition 3.7: Let be $mapping\_object_1$, $mapping\_object_2$ objects of pixel-coordinate, and let be BC_1, BC_2 the corresponding boundary cells. If there exist $Cell_i \in BC_1$, $Cell_k \in BC_2$ and $|x_i - x_k| < D$ and $|y_i - y_k| < D$, where D denotes the calibrating size of the COR or any another size-value, then $mapping\_object_1$ and $mapping\_object_2$ are part of the homogenous object. (see Fig. 3-14).

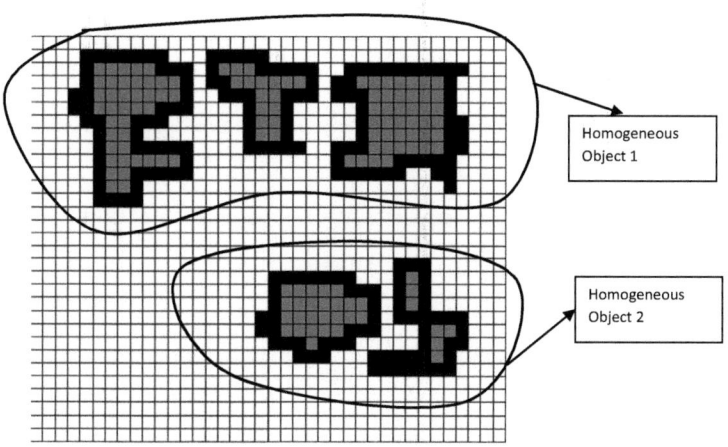

Fig. 3-14 Two Homogeneous Objects

The following pseudo-code algorithm finds all homogenous objects on the environment-map,

//The Runtime of the Algorithm is quadratic, i.e $O(n^2)$.).
IntegerList *Find_All_HomogenousObjects ( )*
{
    1. *Find all Objects, i.e call function FindAllObjects*
    2. For each mapped_object find the neighbor-cells , i.e $\text{Cell}_k \in \mathcal{BC}_i$
    3. For each two mapped_objects and theirs $\mathcal{BC}_1$ *and* $\mathcal{BC}_2$ prove If there exist $\text{Cell}_i \in \mathcal{BC}_1$, $\text{Cell}_k \in \mathcal{BC}_2$ and $|x_i - x_k| < D$ and $|y_i - y_k| < D$
    4. If true then add $\mathcal{BC}_1$, $\mathcal{BC}_2$ into homogenous object.
    5. Return object
}

The process of mapping calculation consists of three different steps:

1) First step is creating an initial mapping. Within this initial map different mapping-objects (obstacles) are identified and theirs boundary cells (pixels) $\mathcal{BC}_i(\text{mapping\_object}_x) \equiv \text{cell}_i$ are calculated. The mapping-objects as such, as well the pixels-coordinate of these boundary cells, are stored.

As the COR has no information about working environments and as there is a SLAM (Simultaneously Localization and Map building) problem, then by the combining of the ultrasonic sensors and CCD camera it is possible to create the map only for the environment within certain distance from robot. The mapping begins with the detecting by the ultrasonic sensors, all visible obstacles around the robot, and measurement of their distance.

After certain of time (in msec), using the data from different ultrasonic sensors, the robot finishes the map building, so it is created an initial mapping frame (see Fig. **3-15**(b)). For the outputs of ultrasonic sensors it is used a digital filter to estimate the data of the environment.

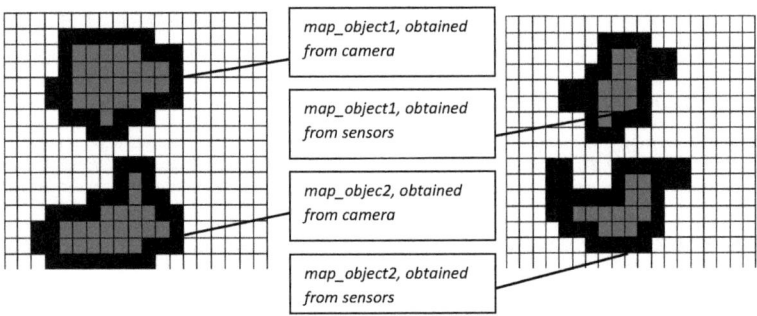

Fig. 3-15 a) Mapping with two objects creates from CCD camera and b) The similar mapping of the same objects created from sonar sensors

The second image plane comes from CCD camera standing over the COR. It is done by openCV or Matlab where using different transformation approaches and using extrinsic and intrinsic camera parameters, it is possible to transform working environments object coordinates into world, camera, image plane and finally pixel coordinates system (Fig. 3-15 (a)).

Using the functions in openCV and Matlab for the segmentation and finding of blobs, it is possible to distinguish different objects in foreground from those in background.

In order to identify uniquely the objects obtained from these two different methods (CCD camera and ultrasonic sensors) within the map, there are applied different new (in this work contribution proposed) approaches. The new approaches are based on the union of all boundary cells of the map-object found on this map; $\bigcup_m^n \mathcal{BC}_i$ (mapping_object$_x$), i.e a merging of the pixels. Finally it is created the unique map of environment. The new created map can be different from the beginnings maps. (see Fig. 3-16).

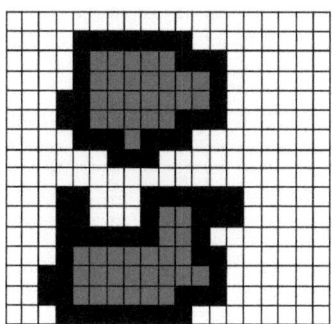

Fig. 3-16 New mapping object as result of fusion of objects from

Fig. 3-15

In Fig. 3-16, Mapping-objects is generated as e result of applying of new approaches over the objects from Fig. 3-15. The approaches are based on the union of the boundary cells between each two objects (the same mapped-objects but from different maps).

2) Second step is creating of the new maps after certain of time (ms), without moving of COR. With the pixels obtained from different camera

frames and the data obtained from fusion of different sensors, the different new maps are created. So comparing of these two or more maps (i.e only the boundary pixels are compared) having all the same coordinate origin, we can distinguish moving objects from static objects. These methods calculate the boundary cells of all objects. If the positions-difference between two objects (same objects on different maps) is bigger than some *error-factor*, which is already calculated, then a moving object is detected.

For the static objects, for each $cell_i \equiv \mathcal{BC}_i(\text{mapping\_object}_x)$ from *map1* and the equivalent $cell_{i\prime} \equiv \mathcal{BC}_{i\prime}(\text{mapping\_object}_x)$ from *map2*, then $Distance(cell_i, cell_{i\prime}) \leq err\_fact$. (see Fig. 3-17).

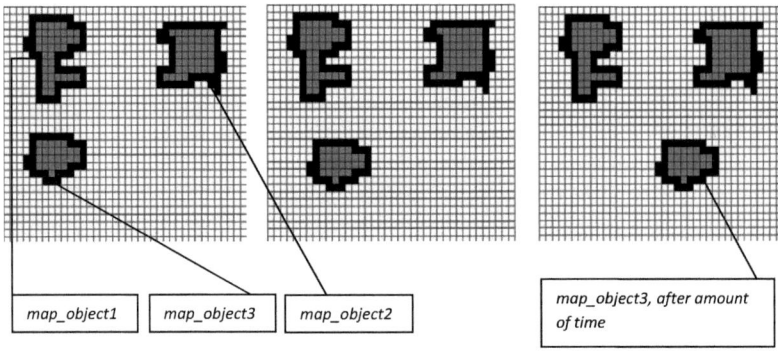

Fig. 3-17 The boundary cells of mapping-object3 after certain are of time.

From Fig. 3-17 it can be seen that only by map_object3 the differences between pixels is un-proportional greater than differences of pixels by the other map_objects. In this way it is determine that map_object3 is a moving object.

Another approach done for this purpose is calculation of camera motion sequences of moving objects from openCV. Comparing of the motion results from camera with results of finding moving objects from ultrasonic sensors should lead to unique identification of moving and static objects within the working environment of COR.

3) Third step is localization of the COR. At the beginning, it is easy to localize the robot position, as the creating of the map derives from the start position of the robot. The new localization of COR is a complex problem and for this purpose it is used the EKF (Extended Kalman Filter) algorithm, as well as new in this work contribution proposed algorithms (exact and heuristic algorithms). Localization problem is described in the next section.

Below it is described the approach in pseudo code.

Algorithm of Mapping of Environment

1. *Create_InitMapping (sens1, sens2,..., sensN, CCD)*
2. *Map = CompareMaps (map1,..., mapN)*
3. *COR_Localalization (Kalman, new algorithms)*

---

1. **Create_InitMapping**
   1.1. *DetMapObjects = Detect all visible obstacles within Work Environment*
   1.2. *Map1 = Transform images from CCD camera into pixel frame*
   1.3. *Map 2 = Apply Algorithms (Digital_Filter (Array of sensors))*
   1.4. *Res_Map = Comapare (Map1, Map2)*
   1.5. *Moving_Objects = Find_AllMovObj(ResMap)*
   1.6. *Static_Objects = Find_AllStatObj(ResMap)*
   1.7. *Update_DataStructure = Store (Moving_Objects, Static_Objects)*, where an Object is a class with information, i.e all cells $cell_i$, $cell_j$, so that $cell_i$, $cell_j \in Object$, $DR(cell_i, cell_j)$. and all $BC_i(mapping\_object_x)$.

COR_Localization

1. *If (Init_Mapping)*
      then *COR_Posit=Start_Position*
   Else
   Cor_Posit = Exact_Heuristics (New Algorithms) and Kalmam Filter. Algortihm

3.3 Localization of Mobile Robot

In case that no globally accurate positioning system is available as for example GPS, the problem of localization became one of the key issues for COR. In chapter 2 different localization approaches are described.

The first main approach already discussed is a passive robot localization where different methods can be used, such as Kalman filter, Map matching, Markov Localization, Monte Carlo Localization, etc.

The second main approach is an active robot localization where different methods such as heuristic search algorithms like D^*, A^* or LRTA^* (Stenz, 1995), (Bulitko et al., 2007) etc, or probabilistic methods such as POMDB-based planning (Gonzales, 2008)

At the navigation with already created maps, two different localization forms can be distinguished, absolute localization and incremental localization. The absolute localization is based on the perception-information and the database, the incremental localization is based on geometrical localization, topological representation of space or localization based on landmark detections.

At the navigation based on map creation, the perception (sensor data fusion and camera information) of the robot during the navigation is proposed to create a map simultaneously, and in respect to the map, the localization of the robot can be calculated, i.e SLAM methods.

Finally at the navigation without maps uses feature environmental information, for example doors, walls etc. The main issues here are the creating of the database with images found in the work environment, or detecting of the landmarks. There is no exact localization of the robot necessary, but the orientation is done in respect to the recognized object.

In this work, a new method, SLAM – Simultaneous Localization and Mapping is presented. In this method the COR during the moving receive information from various sensor, merge all sensor measurements and build a map. The process of map-building is already described in previous section. Finally the Extended Kalman Filter and Neuronal Network approach is proposed, in order to update position beliefs of the COR. In general the SLAM method proposed here is based on the following steps:

1. Perception process, i.e the process is in function of no-visual sensor fusion and CCD camera information. The most advance methods for the sensor fusion like Kalman Filter are proposed. Due to this perception information, the initial position of COR is assumed to be known. In this case the robots uncertainty is small i.e. only "position tracking" should be calculated and no "Global" localization is needed.

2. Map creating process i.e grid map of the environment is generated. By the update of the localization, in case that COR is near to the frame of its generated-map, the new map-calculations should be made. The updating of the map is in function of better positioning belief of the COR.

3. Extended Kalman Filter method - EKF. This method updates beliefs of the COR positioning due to the new gathered information. EKF uses in general two updating methods. The first method made possible the approximation of the COR position in respect to the COR moving (Motion model). The second method made possible the approximation of the COR position in respect to the COR sensors information.

4. Artificial Neuronal Network implementation (ANN). The ANN implemented in this work is a three layer Back-Propagation Neural Network. The network tries to improve the approximated COR positions obtained during the EKF calculations.

    In this section are represented only the main steps of the implementation and the Pseudo-Code algorithm of the ANN. An implementation of the Algorithm in C# and Matlab can be found at chapter 5 (Software Implementation and Simulation results).

The whole process of localization of the COR, is described below.

From Fig. 3-18, follows that, the main part of approach responsible for COR localizing is perception, i.e. continuously merging of all sensors data. The information obtained from perception will be used either to build the map, or to approximate it's already belief position of the COR.
There are two different possibilities distinguished:
- If the COR has no information about its localization, then the creating of the map is necessary, i.e point 2. After the map is created, it is easy to calculate the initial position of the COR. The process of map-calculations is described in previous section.
- If the COR already knows its initial position and the new position after certain of the time should be calculated. In this case, data will be used as input data for EKF , i.e. point 3. After updating of the beliefs position by the EKF approach, the positions will be recalculated by ANN approach. The ANN tries to minimize the difference between the current estimated position of the COR and target position of it.

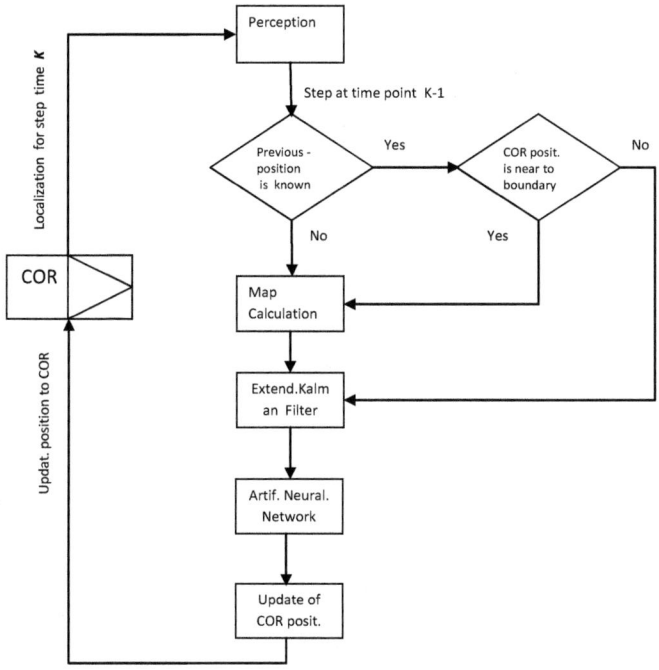

Fig. 3-18 Block - Diagram of COR localization processes

As already mentioned in order to find the previous position it is necessary to create a map.

But also in case the COR position is near to the map frame (see Fig. 3-19), a new map creation is necessary. The new creation of the map, is actually the same process as where the robot had no initial position known. The process of map-calculation is already described in the previous section, and no further instructions in this section are done.

Fig. 3-19 shows that EKF approach is used in order to improve and update the beliefs positions of the COR. Before the technique of using of EKF is described, first it is given a short introduction in to the mobile robot model. There is also given a mathematical model of the function which describes how the robot configuration change in respect to the control inputs in different time sequences.

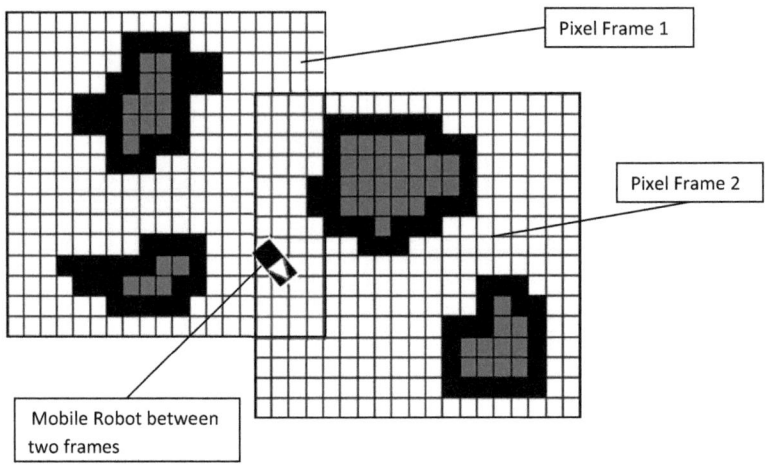

Fig. 3-19 Second pixel frame created

The mathematical models and formulas below (3.1-3.11) are detailed introduced and can be found at (Corke, 2011). In this section only the main equations with relevance to this work, are introduced.

In Fig. 3-20 is given the model of the COR represented by the world coordinate frame [6] and the COR coordinate frame. The COR coordinate frame has x_v-axis in the COR forward direction, with the origin at the center of the COR, and its y_v - axis which increases similar to Y – axis of world coordinate frame. The velocity v of the COR is in the x_v direction. The first focus of interesting in a discrete time model, i.e. behavior of the COR in the discrete time values, and how the configuration of the COR is changed in respect to its control inputs.

---

[6] Due to the simplification of the representing of the COR, no pixel coordinate system is used.

Formally, how to find the function $x\langle k+1\rangle = f(x\langle k\rangle, \delta\langle k\rangle, v\langle k\rangle)$ *, where* $\delta\langle k\rangle = \langle \delta_d, \delta_\theta \rangle$ *represents travelled distance,* δ_d *is the odometer measurement and* δ_θ *is any random measurement noise and* k *is the time step.*

The value $x\langle k\rangle$ *is defined as a true but unknown position of the COR at the time step* k, $\hat{x}\langle k\rangle$ *is denoted as best possible estimate position of the COR at the time step* k, $u\langle k\rangle$ *represents the command sent to the COR at the time step* k.

Odometer measures the distance of the robot from the position at time k *to the position at the time* $k+1$. *In order to define a mathematical model of EKF the Robot model within World-Coordinate system is presented. (Fig. 3-20).*

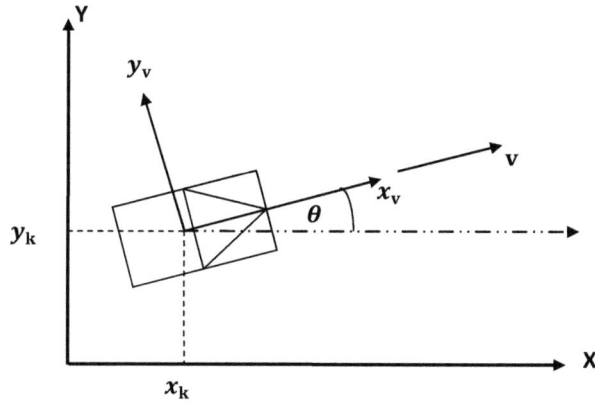

Fig. 3-20 World coordinates and robot coordinates

From Fig. 3-20 the position of the COR can be estimated as (Corke, 2011)

$$\xi\langle k\rangle \sim \begin{bmatrix} \cos\theta\langle k\rangle & -\sin\theta\langle k\rangle & x\langle k\rangle \\ \sin\theta\langle k\rangle & \cos\theta\langle k\rangle & y\langle k\rangle \\ 0 & 0 & 1 \end{bmatrix} \qquad (3.1)$$

, where ξ *is denoted as relative pose of the frame with respect to a reference coordinate frame*

It is assumed that robot moves forward in robots x-direction by δ_d, and then rotate by δ_θ, so for time-step $k+1$ follows:

$$\xi\langle k+1\rangle \sim \begin{bmatrix} \cos\theta\langle k\rangle & -\sin\theta\langle k\rangle & x\langle k\rangle \\ \sin\theta\langle k\rangle & \cos\theta\langle k\rangle & y\langle k\rangle \\ 0 & 0 & 1 \end{bmatrix} \begin{bmatrix} 1 & 0 & \delta_d \\ 0 & 1 & 0 \\ 0 & 0 & 1 \end{bmatrix} \begin{bmatrix} \cos\theta\langle k\rangle & -\sin\theta\langle k\rangle & 0 \\ \sin\theta\langle k\rangle & \cos\theta\langle k\rangle & 0 \\ 0 & 0 & 1 \end{bmatrix} \quad (3.2)$$

$$\xi\langle k+1\rangle \sim \begin{bmatrix} \cos(\theta\langle k\rangle + \delta_d) & -\sin(\theta\langle k\rangle + \delta_d) & x\langle k\rangle + \delta_d \cos\theta\langle k\rangle \\ \sin(\theta\langle k\rangle + \delta_d) & \cos(\theta\langle k\rangle + \delta_d) & y\langle k\rangle + \delta_d \sin\theta\langle k\rangle \\ 0 & 0 & 1 \end{bmatrix} \quad (3.3)$$

or as a 3-vector:

$$\xi\langle k+1\rangle \sim \begin{bmatrix} x\langle k\rangle + \delta_d\langle k\rangle \cos(\theta\langle k\rangle + \delta_d) \\ y\langle k\rangle + \delta_d\langle k\rangle \sin(\theta\langle k\rangle + \delta_d) \\ \theta\langle k\rangle + \delta_d \end{bmatrix} \quad (3.4)$$

Because in reality the odometry cannot be perfect, after adding continuous random variables v_d, and v_θ respectively to δ_d, and δ_θ, can be found the new configuration of the COR at time $k+1$ with respect to the previous configuration at time k and with respect to the odometer, including the odometry error.

$$\xi\langle k+1\rangle \sim \begin{bmatrix} x\langle k\rangle + (\delta_d\langle k\rangle + v_d)\cos(\theta\langle k\rangle + \delta_\theta + v_\theta) \\ y\langle k\rangle + (\delta_d\langle k\rangle + v_d)\sin(\theta\langle k\rangle + \delta_\theta + v_\theta) \\ \theta\langle k\rangle + \delta_\theta + v_\theta \end{bmatrix} \quad (3.5)$$

Because the Kalman Filter are designed for the linear systems and in general the mobile robot is a non-linear system, as first it is shown a linearization of the function $\hat{x}\langle k\rangle$, known as Extended Kalman Filter (EKF), i.e.

$$\hat{x}\langle k+1\rangle = \hat{x}\langle k\rangle + F_x(x\langle k\rangle - \hat{x}\langle k\rangle) + F_v v\langle k\rangle \quad (3.6)$$

where F_x, F_v are Jacobians evaluated from equation (3.5).

The Jacobians are obtained by differentiating equation (3.5) and evaluating them for $v = 0$, giving:

$$F_x = \frac{\partial f}{\partial x}\bigg|_{v=0} = \begin{bmatrix} 1 & 0 & -\delta_d \langle k \rangle - \sin(\theta\langle k \rangle + \delta_\theta) \\ 0 & 1 & \delta_d \langle k \rangle \cos(\theta\langle k \rangle + \delta_\theta) \\ 0 & 0 & 1 \end{bmatrix} \qquad (3.7)$$

and

$$F_v = \frac{\partial f}{\partial v}\bigg|_{v=0} = \begin{bmatrix} \cos(\theta\langle k \rangle + \delta_\theta) & -\delta_d \langle k \rangle \sin(\theta\langle k \rangle + \delta_\theta) \\ \sin(\theta\langle k \rangle + \delta_\theta) & \delta_d \langle k \rangle \cos(\theta\langle k \rangle + \delta_\theta) \\ 0 & 1 \end{bmatrix} \qquad (3.8)$$

Finally, the Extended Kalman Filter prediction equation can be written as:

$$\left.\begin{aligned} \hat{x}\langle k+1|k \rangle &= f(\hat{x}\langle k \rangle, \delta\langle k \rangle, 0) \\ \hat{P}\langle k+1|k \rangle &= F_x\langle k \rangle \hat{P}\langle k|k \rangle F_x\langle k \rangle^T + F_v\langle k \rangle \hat{V} F_v\langle k \rangle^T \end{aligned}\right\} \qquad (3.9)$$

where $\hat{x}\langle k+1|k \rangle$ is the estimation of the $x = (\hat{x}, \hat{y}, \hat{\theta})$ at time $k+1$ with respect to information up to step-time k. The variable $\hat{P}$ is a Covariance Matrix and represents the estimated COR configuration, and finally the variable $\hat{V}$ represent the estimating of the covariance for the odometry noise, and normally this value is not known.

The last equations describe the motion model of the COR, i.e. odometry, but there is also the sensor model of the COR, i.e observation part of the EKF. In (Corke, 2011) a version of sensor model observation can be found. In this work this sensor model is proposed and only the final equations are given.

The below equations are the final equations after updating of the predicted state equations of the COR (3.9).

$$\left.\begin{array}{l}\hat{x}\ \langle k+1\,|\,k+1\rangle = \hat{x}\ \langle k+1\,|\,k\rangle + K\langle k+1\rangle v\langle k+1\rangle \\ \hat{P}\ \langle k+1\,|\,k+1\rangle = \hat{P}\ \langle k+1\,|\,k\rangle F_x\langle k\rangle^T - K\langle k+1\rangle H_x\langle k+1\rangle \hat{P}\ \langle k+1\,|\,k\rangle\end{array}\right\}$$
(3.10)

The equations (3.10) are the Kalman Filter update equations. In these equations for the computing of the values from the time step $k+1$ denoted as $k+1\,|\,k+1$, are used the predicted values $k+1\,|\,k$ and apply information from the step $k+1$. The Kalman Gain Matrix K is defined as follows:

$$\left.\begin{array}{l}S\ \langle k+1\rangle = H_x\ \langle k+1\rangle \hat{P}\ \langle k+1\,|\,k\rangle H_x\langle k+1\rangle^T + H_w\langle k+1\rangle \widehat{W}\ \langle k+1\rangle H_w\langle k+1\rangle^T \\ K\ \langle k+1\rangle = \hat{P}\ \langle k+1\,|\,k\rangle H_x\langle k+1\rangle^T + S\ \langle k+1\rangle^{-1}\end{array}\right\}$$

where $\widehat{W}$ is the estimated covariance of the sensor noise. (3.11)

With (3.11), it is possible to create an estimator that uses COR motion model and sensor observation of the environment. (Corke,2011)

3.3.1 New Back-Propagation Neural Network approach

In this section, it will be described the process of using of the Artificial Neural Network in further minimizing of the errors during COR localization.

It is supposed that COR should move from the already known position at time t_0 to the any target position at time t_n , see (Fig. 3-21). Further it is supposed the trajectory is calculated (see next section), and the target position coordinates are known.

Under the ideal condition, i.e. without any occurring errors, the COR for any discrete time value t_k should be at already calculated position, i.e estimated position of the COR is the same as calculated position from the map. Formally $x\langle k \rangle = \hat{x}\langle k \rangle$. Although for this purpose the EKF approach is used, the error generated from odometry noise and sensor noise can still remain.

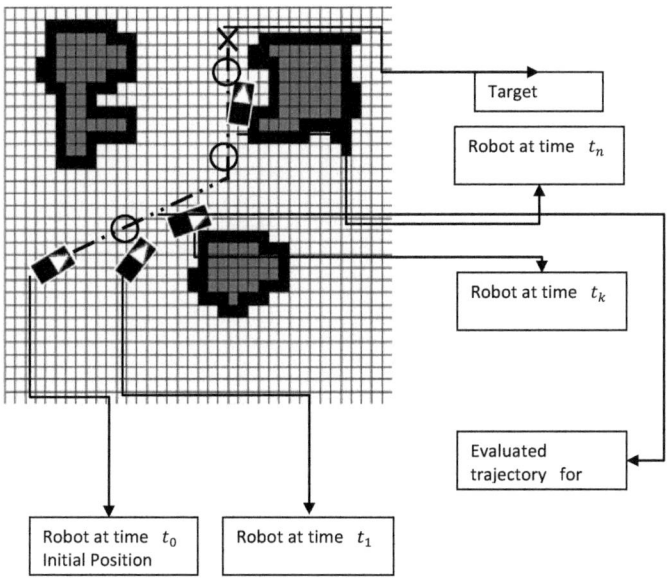

Fig. 3-21 The supposed calculated Trajectory of COR and its estimated positions for different time-steps

The end-result of an already described complicated mathematical model of EKF is an estimated

COR position. The difference between this estimated position and the COR position already calculated[7] in map, in this work is denoted as $\Delta_k = |x\langle k \rangle - \hat{x}\langle k \rangle|$, see Fig. 3-22,[8] where $x\langle k \rangle = \langle x_k, y_k \rangle$, $\hat{x}\langle k \rangle = \langle x_{\hat{k}}, y_{\hat{k}} \rangle$ and $\Delta_k = \sqrt{(x_k^2 - x_{\hat{k}}^2) + (y_k^2 - y_{\hat{k}}^2)}$.

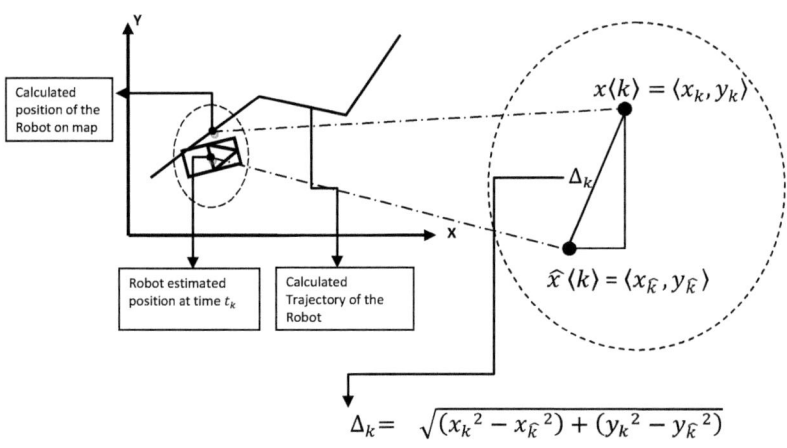

Fig. 3-22 Difference between COR estimated position and position calculated on map

Every non-zero value of Δ_k (Fig. 3-22) for any time-step k has an impact on other Δ_{k+i} for $i \in \{1, 2, \ldots n\}$. This impact in general is not linear function and due to the odometry and sensor noise is very difficult to find an appropriate mathematical model for it.

For this purpose, in this work a three layer Back-Propagation Neural Network is implemented.

This art of Artificial Neural Network learns by examples. If ANN is given some examples of that what should be done ($\Delta_k \xrightarrow{yields} 0$) the algorithm

---

[7] The position of the COR calculated on the map, $x\langle k \rangle = \langle x_k, y_k \rangle$ is used as a true position for the robot

[8] Due to simplification of the mathematical model of distance, no pixel coordinate is used. In the implementation, the converting of the dimensions on the pixel coordinate system is done.

changes the networks weights in order to give the desired output for the particular input. This training of the network then should be applied on all other examples.

In Fig. 3-23 a three layer Back-Propagation Neural Network with 7 neurons is shown. In order to obtain third value from two previous values, a three layer Back-Propagation is enough good. The first layer is the input layer, the middle layer is the hidden layer and finally the third layer is the output layer. The neurons are denoted with N1 (first neuron) up to N7 (last neuron). There are two bias[9] neurons (N1 and N4) which outputs are always 1.

The weights from neurons of one layer to neurons of another layer are denoted as w_{ji}, where j- represents the neuron number of actual layer and i- represents the neuron number of previous layer

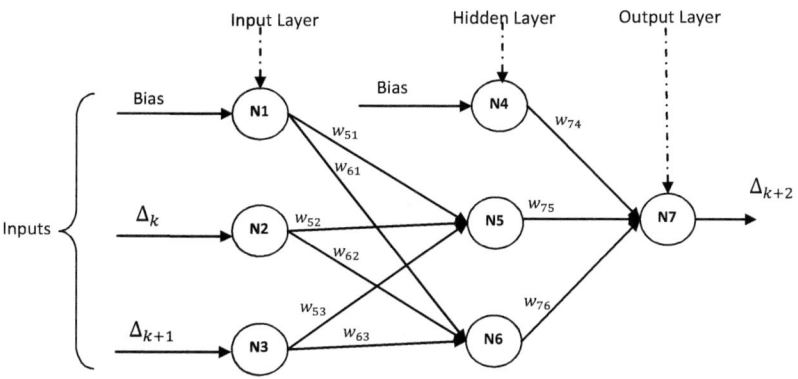

Fig. 3-23 Neuronal schema

The implementation of Back-Propagation Learning Algorithm consists of 5 steps:

---

[9] The bias neurons help the network to learn the patterns. With bias neurons the output of the activation function can be shifted to the left or right on the x-axis.

- Step 1: Is the weight initialization. The weights will be initialized to the random numbers. It is preferred to start with small random numbers.
- Step 2: Is the calculation of output-levels (see Fig. 3-23). There will be distinguished between the output-level of an input neuron and the output-level of one hidden or output neuron.
    - In case of an input neuron the output-level of it is the same as its input value.
    - In case of an output or hidden neuron, the output level can be found by the formula $O_j = \varphi(net_j) = \frac{1}{1+e^{-\alpha \cdot net_j}}$, where φ is the activation function (in this work it is used sigmoid function, see Fig. 3-25), α is a degree of fuzziness, and $net_j = \sum w_{ji} * O_i - \theta_j$. The O_i represents the input at neuron i, and θ_j is the node threshold. A processing unit sums the inputs, and then applies a non-linear activation function (threshold function, i.e. sigmoid function). Output transmits the results to other neurons.

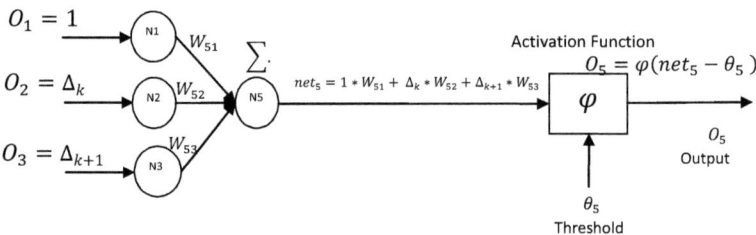

Fig. 3-24 Activation Function and Output calculation Example for Hidden Neuron N5

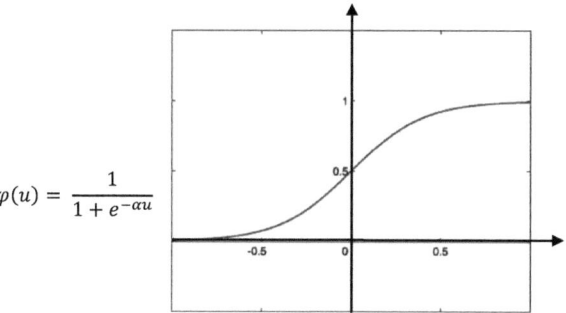

Fig. 3-25 Continuous Log-Sigmoid Function [Matlab]

- Step 3: The training phase of the Weights. In this phase the back-propagation algorithm calculates the output-errors of neurons. There are also two different possibilities distinguished:
    - In case of an output neuron (in our case N7), the output error will be calculated as: $\delta_7 = O_7 * (1 - O_7) * (Target_7 - O_7)$, where O_7 is the actual output activation at output neuron N7 and $Target_7$ is desired output at N7.
    - In case of the hidden neuron, the output error will be calculated as:

    $\delta_j = O_j * (1 - O_j) * \sum \delta_k * W_{kj}$ where δ_k is the output error of the neuron Nk, which is in the following layer (in our case output layer) and which input comes from actual hidden neuron Nj.

    In our case, because there is only one Output neuron N7, then for example the output error for hidden neuron N5 can be

calculated as: $\delta_5 = O_5 * (1 - O_5) * \delta_7 * W_{75}$, (see Fig. 3-26.)

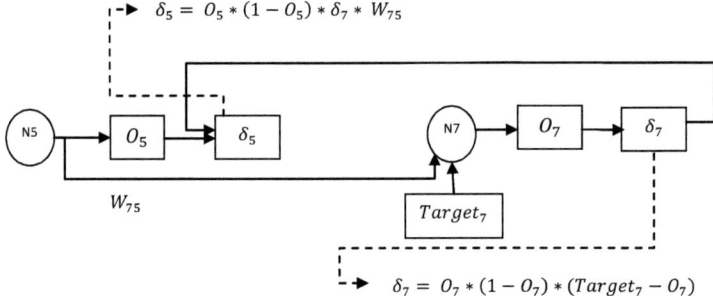

Fig. 3-26 Example of the finding of the output-error for a hidden neuron N5.

In Fig. 3-26 as first there will be calculated the output-level O_7 at the output neuron N7, and output-level O_5 at hidden neuron N5. Then at output neuron N7 will be calculated the error-output δ_7. Finally this value it is used for the calculating (back propagate) for the output-error of hidden neuron N5.

- Step 4: After all output errors are calculated, the next step is the weight-adjustment. The calculating of weight-adjustment will be done using the formula:
$W_{ji}^+ = W_{ji} + \eta * \delta_j * O_i$, where W_{ji}^+ is the weight from neuron Ni to neuron Nj at the next iteration, η is the learning rate ($0 < \eta < 1$) which can change the speed of learning.

- Step5: Perform the next iterations for the new instances (if there exists more test-examples) of the inputs and repeat process from step 2. Finally repeat the process until the error criteria is satisfied. (see Fig.3-27)

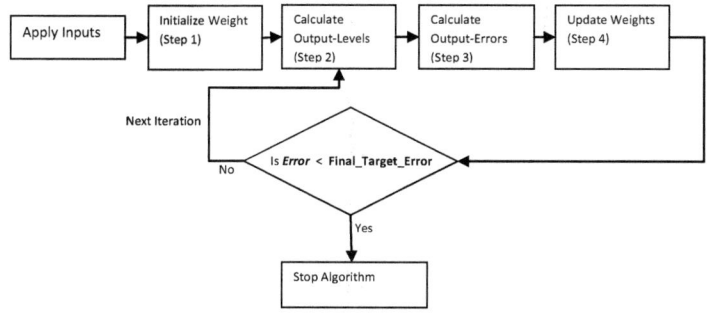

Fig. 3-27 Overview of the Back-Propagation Algorithm

Below (see also Fig. 3-23 and Fig. 3-27), the Back-propagation algorithm for the network-solution is presented.

There is a three layer Back-Propagation Neural Network with one input-layer, one hidden-layer, one output-layer and 7 neurons: 3 inputs-neurons (N1,N2,N3 , where N1 is Bias-neuron), 3 hidden-neurons (N4,N5 ,N6, where N4 is Bias-neuron) and 1 output-neuron N7.

The training instance is given as follows: The inputs at Input-neurons are: Input1 = Bias = 1, Input2 = Δ_k , Input3 = Δ_{k+1}

1. The weights are initialized as follows: Notice that no weight to one Bias-neuron is necessary

    1.1. From Input-neurons to Hidden-neurons:

    From Input – Bias (N1) to Hidden-neuron (N5) : W_{51}

    From Input – Bias (N1) to Hidden-neuron (N6) : W_{61}

    From Input – neuron (N2) to Hidden-neuron (N5) : W_{52}

    From Input – neuron (N2) to Hidden-neuron (N6) : W_{62}

    From Input – neuron (N3) to Hidden-neuron (N5) : W_{53}

    From Input – neuron (N3) to Hidden-neuron (N6) : W_{63}

    1.2. From Hidden-neurons to Output-neuron:

    From Hidden – Bias (N4) to Output-neuron (N7) : W_{74}

    From Hidden – neuron (N5) to Output-neuron (N7) : W_{75}

From Hidden – neuron (N6) to Output-neuron (N7) : W_{76}

2. The calculation of the output levels (O_i represents the Output of the neuron i). As first it will be calculated the output-level of the Input-neurons, then output-level of the hidden neurons, and finally the output-level of the output-neuron: Notice that the output-level of the Bias neuron is equally to 1. The variables α is a constant value and θ_j is a node threshold.

   2.1. The Output-Level for Input-neurons, is the same as the input-value,

   i.e. $O_1 = Bias = 1$, $O_2 = \Delta_k$, $O_3 = \Delta_{k+1}$

   2.2. The Output-Level for the hidden-neurons:

   For Hidden-Bias neuron N4 : $O_4 = Bias = 1$

   For Hidden-neuron N5 : $O_5 = \dfrac{1}{1+e^{-\alpha *(1*w_{51}+ \Delta_k*w_{52}+ \Delta_{k+1}*w_{53} -\theta_5)}}$

   For Hidden-neuron N6 : $O_6 = \dfrac{1}{1+e^{-\alpha *(1*w_{61}+ \Delta_k*w_{62}+ \Delta_{k+1}*w_{63} -\theta_6)}}$

   2.3. The Output-Level for the Output-neuron :

   $O_7 = \dfrac{1}{1+e^{-\alpha *(1*w_{74}+ O_5*w_{75}+ O_6*w_{76} -\theta_7)}}$

3. The Calculation of the output-errors (δ_i) of the neurons : The process goes backwards, i.e as first the output-error of output-neuron will be calculated, and then the output-error of the hidden-neurons: Notice that $Target_7$ denoted the desired output at output-neuron.

   3.1. For Output–neuron N7 : $\delta_7 = O_7 * (1 - O_7) * (Target_7 - O_7)$

   3.2. For Hidden-neuron N4: $\delta_4 = O_4 * (1 - O_4) * \delta_7 * W_{74}$

   For Hidden-neuron N5: $\delta_5 = O_5 * (1 - O_5) * \delta_7 * W_{75}$

   For Hidden-neuron N6: $\delta_6 = O_6 * (1 - O_6) * \delta_7 * W_{76}$

4. The Weight-Adjustment Calculation. Notice that η is the learning rate ($0 < \eta < 1$) which can change the speed of learning.

   4.1. From Input-neurons to Hidden-neurons :

From Input – Bias (N1) to Hidden-neuron (N5) : $W_{51}^+ = W_{51} + \eta * \delta_5 * O_1$

From Input – Bias (N1) to Hidden-neuron (N6) : $W_{61}^+ = W_{61} + \eta * \delta_6 * O_1$

From Input – neuron (N2) to Hidden-neuron (N5) : $W_{52}^+ = W_{52} + \eta * \delta_5 * O_2$

From Input – neuron (N2) to Hidden-neuron (N6) : $W_{62}^+ = W_{61} + \eta * \delta_6 * O_2$

From Input – neuron (N3) to Hidden-neuron (N5) : $W_{53}^+ = W_{53} + \eta * \delta_5 * O_3$

From Input – neuron (N3) to Hidden-neuron (N6) : $W_{63}^+ = W_{63} + \eta * \delta_6 * O_3$

4.2. From Hidden-neurons to Output-neuron:

From Hidden – Bias (N4) to Output-neuron (N7) : $W_{74}^+ = W_{74} + \eta * \delta_7 * O_4$

From Hidden – neuron (N5) to Output-neuron (N7) : $W_{75}^+ = W_{75} + \eta * \delta_7 * O_5$

From Hidden – neuron (N6) to Output-neuron (N7) : $W_{76}^+ = W_{76} + \eta * \delta_7 * O_6$

For the next iteration the new Weights from the programming view can be obtain from $W_{ji} = W_{ji}^+$, and the process repeated until $|Target_7 - O_7| \leq \varepsilon$. In ideal case $\varepsilon = 0$.

After the training of an instance, the new instance of values can be added, in our case as inputs will be: Input1 = Bias = 1, Input2 = Δ_{k+1} , Input3 = Δ_{k+2}.

The problem at neuronal-network approaches is the small number of the training instances, i.e as less number of the test examples as higher possibility of the calculation error.

In case of our problem in order to neutralize the possibility of error obtained from ANN, the new weight function is introduced.

Let suppose for the calculated trajectory of COR from Start-Position to Target-Position the N time-step calculation are needed. The new approximate position $\hat{x}\langle k \rangle$ then is :

$$\hat{x}\langle k \rangle = f(N, k, \Delta_k) = \begin{cases} \hat{x}\langle k \rangle \pm \frac{2*k}{N} * \Delta_k \text{ , If } 0 \leq k \leq \frac{N}{2} \\ \hat{x}\langle k \rangle \pm \Delta_k \text{ , else} \end{cases}$$

where Δ_k is the error obtained from ANN.

For small k, the number of test-instances for Neural-Network approach was small, i.e the value Δ_k obtained in this way is not so reliable, therefore only $\frac{2*k}{N} * \Delta_k$ is used.
For enough number of test examples for ANN, i.e large k, the obtained value Δ_k can be used as fairly value.

3.3.2 Summary

In section 3.3 a new approach for localization problem is proposed (Fig. 3-18). In this new approach two state of the art methods like Extended Kalman Filter (EKF) and Artificial Neural Network are used. Each of this method calculates the errors and updates beliefs of the robot positioning. During the EKF to create an estimator use a complex mathematical model (based on motion model and sensor observation), the ANN in this work as inputs use only two last already estimated positions derivate from EKF.

The new position predicted after applying of ANN is exacter value than before applying. Because the ANN depends from the number of training instances, the values obtained from ANN at the beginning are not enough reliable. Only after certain number of trainings instances the ANN values are taken proportionally in consideration.

3.4 Path Planning and Navigation

The navigation problem and path planning are dealing with the finding of the shortest, sometimes optimal way for robot moving towards to any target position. The obstacle avoidance as well as trajectory following are the main part of the robot navigation problem. For this purpose, in the literature can be found the different studies and approaches (Ito, 2009). For example using of the potential fields that satisfy stability in Lyapunov sense, or using of Model Predictive Control (MPC) on short prediction horizon can be found in case of short prediction horizons, or at least new computer vision approaches.

In the last years the vision sensing concepts and approaches for the mobile robotics are developed (Ito, 2009). Using these methods and approaches the Mobile Robot Navigation can be classified as Indoor or Outdoor Navigation. The indoor navigation approaches can be further classified as (Ito, 2009), (see chapter 2).

- *Navigation with maps* - includes geometric environments information

- *Navigation based on map creation* – sensors are used to create the build the environment map
- *Navigation without map* – do not use any environment representation but merely recognized objects that can be tracked

The outdoor navigation include landmark and obstacle detection, map creation, actualization and position estimation. The outdoor navigation can be further classifies as:

- *Structured outdoor navigation* – For example roads are interesting research area
- *Unstructured outdoor navigation* – mainly based on environmental camera perception

In all these methods, the avoiding of obstacles should be done during navigation towards the goal position. This can be done by using either incremental (the initial position of the robot is already known) or absolute localization (in case that initial robot position is unknown) (Ito, 2009). Both methods can use different landmarks and analyze its position on map such as an estimating of robot position can be done; ore can use also different techniques like Kalman Filters, or Monte Carlo Localization (MCL) etc. Once it is known the new estimated robot positions, the robot can move from actual to the new position. The localization techniques and approaches are in general introduced in chapter 2. The localization approaches proposed in this work can be found in section 3.3 (Localization of Mobile Robot).

In case that no prior information about the environment and no map is available, then the robot position should be estimated and the map representation of environment should be calculated. The navigation approaches based on these methods known as *Navigation based on map creation,* are calculation-expensive. In this methods the robots perception during the navigation can be used to create a map. The quality and the performance of such created maps depend from the preciosity of the robots sensor system, sensors fusion or even by the using of the vision system. Then by using of the localization methods on the created map, as previously discussed, the robot can navigate from actual to new position towards the target position. The proposed approach of map creating for this work can be found at chapter 3.2 (Map calculating of the environment). The general information about map creation can be found at chapter 2 (State of the art).

At navigation without maps, without any information about environment and where no map can be created, the different techniques are used. They are based mainly on different features of environment like, doors, walls or any another landmarks. Some of the significant strategies are (Ito, 2009) :

- *Navigation based on Optical Flow* – for example using of optical perspective view and optical flow perception, as well as object tracking
- *Navigation based on database object recognition* – creating of database with images featured from work environment
- *Navigation based on object recognition without database* - mainly based on the landmark detection.

Some of the simple class of robots using this art of navigation, known also as *reactive navigation* (Corke, 2011), are *Breitenberg* vehicles or *Bugs*- simple automata robots (based on several known bug-algorithms, for example bug2 algorithms). An Introduction on this art of navigation can be found at Chapter 3 (State of the art). For more researching on the different navigation approaches without maps can be found at (Nakamura, Asada, 1995).

In this work is treated the navigation problem of an indoor mobile robot where neither prior information about the working environment nor initial COR position is known. The solution proposed for this kind of navigation is the self-localization of the COR and creating of the grid map as representation of the working environment.

The creating of the map in this work is described in section 3.1 and 3.2. There is proposed the using of the vision system and grid decomposition named *occupancy grid*. Several new approaches in this work about the detection and manipulation with objects on this greed are described.

In section (3.4) two new approaches for path planning on the occupancy grid map are proposed and implemented. The approaches are based on collision detection and path generating. The new heuristic approach proposed in section 3.4.1 uses the representing of the obstacles on the occupancy greed, where the occupied cell greed is part of the obstacle. Finding the geometrical intersection (by using the linear interpolation) between COR position and the obstacle it is possible to avoid the collision and to generate a path.

In section 3.4.2 is proposed a new approach based on *hypergraph partitioning* and *hypertree decomposition*. Representing only the free-cells on hypergraphs and then using the hypertree decomposition approach (Dermaku et al.,2007), can be generated the free paths from start to target position. In case of existing of such paths, this method can provide an optimal one.

There exists also possibility of mix of approach based on geometrical intersection and approach based on hypertree decomposition. There can be used additionally a heuristics by the decision during the path segment generation. The heuristic try to find the path through the lower density of obstacles on map toward to target position.

3.4.1 A new heuristic algorithm for path planning based on geometrical intersection [10]

The optimal path is good, but sometimes is run-time and memory consuming, especially on the large areas of the COR working environment. In those cases heuristic approaches can be used in order to find a solution. The better planned heuristic leads to the near optimal path, but in general the heuristic do not guaranty the optimal solution. The heuristic approaches try to find the good solution in responsible time, and usually can be used when the optimality can be sacrificed for runtime. In literature can be found the best known heuristic algorithms on graphs developed for this purpose, like , static search algorithm A^* (Hart, Nilsson, Raphael, 1968) ,dynamic re-planning-algorithms D^* (Stentz, 1994), learning real time $LRTA^*$ (Korf, 1990) or its extension algorithm named Path-Refinement Learning Real-time Search Algorithm (Bulitko at.al , 2007) , etc.

In this work, a new heuristic algorithm based on geometrical intersection is presented. The algorithm in general tends to find the different sets with neighbor cells (of any calculated cell) that can be connected to each other without calculation of all between cells. The advantage of such an approach is that only several areas with cells in greed-map $\mathcal{M}$ will be searched. The disadvantage is that in case the obstacles have very complex form, not always in this way can be found the connection between cell-areas. In such cases the algorithm uses additionally the hypertree decomposition approaches (see chapter 3.4.2). It can be shown that in some cases the algorithm can produce solutions near to optimal one.

---

[10] The algorithm is partially published at (A.Dermaku, 2013)

Let suppose there exist an *occupancy grid* map as proposed in section 3.2. This map is created by using of the vision system (see section 3.1) and applying of the cell decomposition methods (see section 3.2). Basically the environment is divided into cells (pixels) of equal size $cell_i = \{x_i, y_i, F_i\}$ (chapter 3.2). The values x_i, y_i represents the coordinate values of Pixel Coordinate System and F_i can be *True* or *False* and denotes if a cell is occupied or not. In this work, the black filled cell means the cell is occupied (Fig. 3-28). The map is represented as a set of the elements (cells) :

$$\mathcal{M} = \bigcup_{i=1}^{N} cell_i \ , i \in \{1,2,\dots,N\} \tag{3.12}$$

in pixel coordinate system where $N = x_{max} * y_{max}$. Values x_{max}, y_{max} denote the maximal coordinate values in x and y axe.

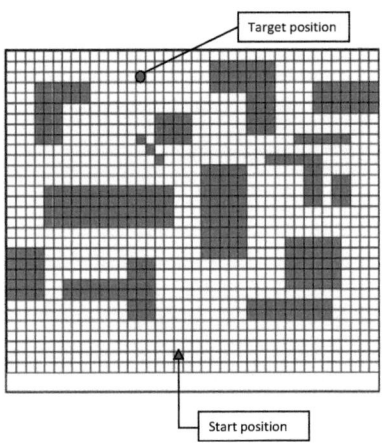

Fig. 3-28 Binary Occupancy grid map

For $cell_i$ the relation between cell-index i and cell-coordinate values x_i, y_i can be calculated from equations $x_i = i \bmod x_{max}$ and $y_i = i / x_{max}$, where, $i = y_i * x_{max} + x_i$ (see also chapter 3.2 and Fig. 3-7).

In this work, the COR is represented by a triangle mask in the greed map within a greed cell, and its size is assumed to be equally to the size of a cell. In general there are several possibilities for representing of the COR. One possibility is to divide environment into squared cells of the size equally to the COR (as in this work is done). The advantage of this art of representation is the less calculation steps, especially by the calculation of the safety distance between COR and obstacle. Other possibilities are to divide the environment independently from the COR size, and then all calculation to adopt in respect to the COR size. For the different possibilities see chapter 3.2 and chapter 5 (Software Implementation and Simulation results).

For further terms used in this section, like *neighbor-cells*, *continually-connected* cells, *direct-connected* cells or *homogenous objects* see *Definitions* in chapter 3.2.

As first it is described the algorithm for avoiding of one object from COR. The cases where the objects are closed convex polygons and where COR and Target Positions are outside of the object, are very easy cases. For the algorithm described below in this work, these cases are trivial.

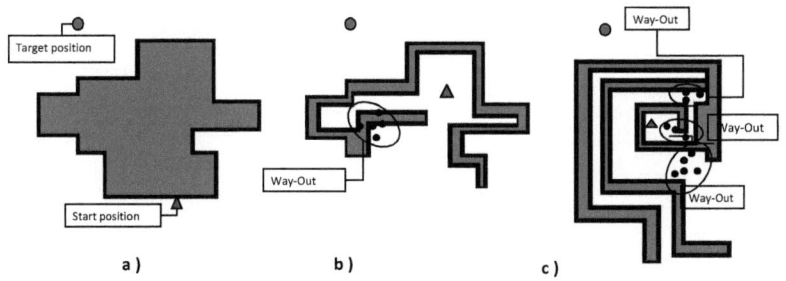

Fig. 3-29 Some of possible positions between COR, Target Positions and Concave Object

The problematic examples are the open formed concave objects where the COR, Target Positions ore both are inside the object. Some of these cases can be found at Fig. 3-29. From Fig. it can be seen that the easiest case is (a) because of the closed concave form of object and because both COR and Target positions are outside of the object. The case (b) is more difficult because the object is not closed and the COR is inside the object. Especially the region "way out" between COR and outside is most problematic. On such difficult regions, in case the algorithm finds no path, can be applied method hypertree decomposition (see section 3.4.2). The case (c) is the most difficult case because of the labyrinth forms of the object and because the COR is inside the object. Similar to the case (b), here can be applied additionally the hypertree decomposition method.

Below the procedure of the avoiding of one object from the COR is described. In this work the COR Start position is named as SP and Target position is named as TP. Due to the similarity it is used the case where SP and TP are outside the object.

For the object (Fig. 3-30), depending on the relative position between SP with coordinate values X_{SP}, Y_{SP} and TP with coordinate values X_{TP}, Y_{TP}, in order to avoid this object the algorithm can find two different areas with cells (left or right). Let be the most-left cell of the object (further named as ML) with the respective X and Y coordinate values X_{ML}, Y_{ML} and the most-right cell of the object (further named as MR) with the respective X and Y coordinate values X_{MR}, Y_{MR}.

It follows that :

$$\left. \begin{aligned} \Delta_{X(TP,MR)} &= X_{MR} - X_{TP}, \\ \Delta_{X(SP,MR)} &= X_{MR} - X_{SP}, \\ \Delta_{X(MR,ML)} &= X_{MR} - X_{ML}. \end{aligned} \right\} \quad (3.13)$$

In case that $\Delta_{X(TP,MR)} + \Delta_{X(SP,MR)} < \Delta_{X(MR,ML)}$ the algorithm will choose the right-side avoiding, otherwise a left-side avoiding will be chosen. (Fig. 3-30)

From (equation 3.13) follows that if $X_{MR} < X_{TP} + X_{SP} - X_{ML}$ the right-side will be chosen, otherwise the left-side will be chosen.

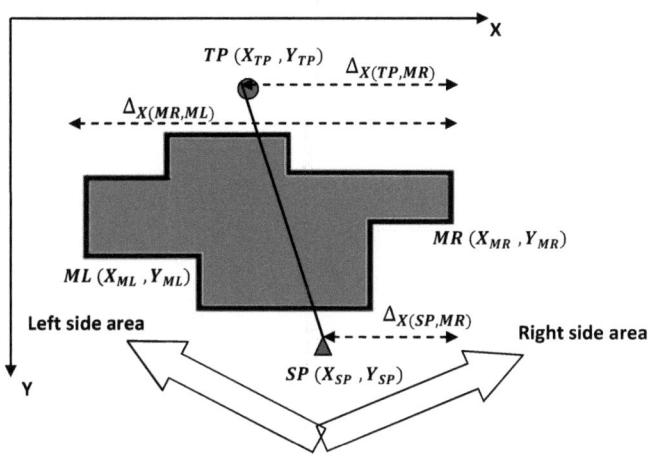

Fig. 3-30 Decision for left or right avoiding

In the next step the algorithm finds the intermediate-positions from SP to TP.

1. For left area: Finds the most-left cell of the object (ML), most-down cell between ML and SP (further named as MDL) and most-upper cell between ML and TP (further named as MUL).
2. For right area: Finds the most-right cell of the object (MR), most-down cell between MR and SP (further named as MDR) and most-upper cell between MR and TP (further named as MUR).

Please note that MDL, MUL, MDR and MUR should be calculated only in case there is no connection possible between ML and TP, ML and SP, MR and TP, MR and SP respectively.
- Continue recursively:
  - Between ML and MDL as well as between MDL and SP find all new most-down cells MDL' until connection achieved
  - Between ML and MUL as well as between MUL and TP find all new most-upper cells MUL' until connection achieved
  - Between MR and MDR as well as between MDR and SP find all new most-down cells MDR' until connection achieved
  - Between MR and MUR as well as between MUR and TP find all new most-upper cells MUR' until connection achieved

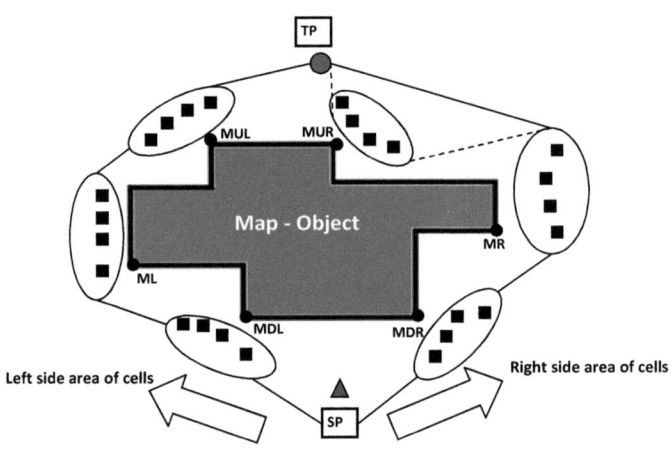

Fig. 3-31 The left and right between-cells during avoiding of the Object

Below at Alg. 4.1 is given a pseudo-code for finding all Down Cells of object with relevance for the proposed algorithm in this work. (see above point 2.). In case that left down points are required, the start parameter P can be ML otherwise P is MR.

Algorithm 3.1 : Most_Down (obj, P , SP, left)

  // Finds recursively all Most_Down Cells of Object obj, between P and SP

1. *if Conenct(P,SP)*
2.   *terminate and return*
3. *end*
4. *If (left)*
5. $p_1 \leftarrow obj.MDL'$
6. *else*
7. $p_1 \leftarrow obj.MDR'$
8. *end*
9. $list_p \leftarrow list_p.Add(p_1)$
10. **Most_Down (obj, P, p_1)**   //recursive approach
11. **Most_Down(obj, p_1, SP)**   //recursive approach

Below at Alg. 3.2 is given a pseudo-code for finding all Upper Cells of object with relevance for the proposed algorithm in this work. (see above point 2.). In case that left upper points are required, the start parameter P can be ML otherwise P is MR.

Algorithm 3.2 : Most_Upper (obj, P,TP,left)

  // Finds recursively all Most_Upper Cells of Object obj, between P and TP

1. *if Conenct(P,TP)*
2.   *terminate and return*
3. *end*
4. *If (left)*
5. $p_1 \leftarrow obj.MUL'$
6. *else*
7. $p_1 \leftarrow obj.MUR'$
8. *end*
9. $list_p \leftarrow list_p.Add(p_1)$
10. **Most_Upper (obj, P, p_1)**   //recursive approach
11. **Most_Upper(obj, p_1, TP)**   //recursive approach

- For all cells (ML,MDL, *MDL'*, MUL, *MUL'*,MR,MDR , *MDR'*, MUR, *MUR'*) find the neighbor cells that are not *directly-connected i.e.* all not filled neighbor cells . There are eight different cases distinguished :
    - Case 1: If both, SP and TP are inside the object, and algorithm calculates cells between ML and TP, as neighbor cells will be choose the *right not directly-connected neighbor cells* as well as so-called *Upper Left-Right cells* (Fig. 3-34)
    - Case 2: If both, SP and TP are inside the object, and algorithm calculates cells between ML and SP, as neighbor cells will be choose the *right not directly-connected neighbor cells* as well as so-called *Down Left-Right cells*
    - Case 3: If both, SP and TP are inside the object, and algorithm calculates cells between MR and TP, as neighbor cells will be choose the *left not directly-connected neighbor cells* as well as so-called *Upper Left-Right cells*
    - Case 4: If both, SP and TP are inside the object, and algorithm calculates cells between MR and SP, as neighbor cells will be choose the *left not directly-connected neighbor cells* as well as so-called *Down Left-Right cells*
    - Case 5: Is at least one of them (SP or TP) outside the object, and algorithm calculates cells between ML and TP, as neighbor cells will be choose the *left not directly-connected neighbor cells* as well as so-called *Upper Left-Right cells* (Fig. 3-34 .a)
    - Case 6: Is at least one of them (SP or TP) outside the object, and algorithm calculates cells between ML and SP, as neighbor cells will be choose the *left not directly-connected neighbor cells* as well as so-called *Down Left-Right cells*
    - Case 7: Is at least one of them (SP or TP) outside the object, and algorithm calculates cells between MR and TP, as neighbor cells

will be choose the *right not directly-connected neighbor cells* as well as so-called *Upper Left-Right cells*
- Case 8: Is at least one of them (SP or TP) outside the object, and algorithm calculates cells between MR and SP, as neighbor cells will be choose the *right not directly-connected neighbor cells* as well as so-called *Down Left-Right cells*

Below at Alg. 3.3 is given a pseudo-code for finding all neighbor cells that are not *directly-connected,* as already explained above.
- The parameter P is any cell calculated from Algorithm. 3.2.
- *inside* is Boolean variable (true mean SP and TP are inside object, false mean at least one of them is outside of object).
- *left* is Boolean variable (true mean P is any most-left cell of object, false mean P is any most-right cell of object),
- *down* is Boolean variable (true mean P is any most-down cell of object, false mean P is any most-upper cell of object).

Algorithm 3.3 : $Not\_\mathcal{D}_c\_Neighbor\ (P, inside, left, down)$

1. $for\ \forall\ Cell_i\ |\ \mathbb{N}(cell_i,\ P) \land F_i = False$ // Not filled neighbor cells of cell P
2. $If\ (\ inside\ \&\ left\ \&\ \neg down\ \&\ (X_i > X_p)\ )$ // Case 1
3. $list_p \leftarrow list_p.Add(Cell_i)$
4.   else
5. $If\ (\ inside\ \&\ left\ \&\ down\ \&\ (X_i > X_p)\ )$ // Case 2
6. $list_p \leftarrow list_p.Add(Cell_i)$
7.   else
8. $If\ (\ inside\ \&\ \neg left\ \&\ \neg down\ \&\ (X_i < X_p)\ )$ // Case 3
9. $list_p \leftarrow list_p.Add(Cell_i)$
10.   else
11. $If\ (\ inside\ \&\ \neg left\ \&\ down\ \&\ (X_i < X_p)\ )$ // Case 4
12. $list_p \leftarrow list_p.Add(Cell_i)$
13.   else
14. $If\ (\neg\ inside\ \&\ left\ \&\ \neg down\ \&\ (X_i > X_p)\ )$ // Case 5
15. $list_p \leftarrow list_p.Add(Cell_i)$
16.   else
17. $If\ (\neg\ inside\ \&\ left\ \&\ down\ \&\ (X_i > X_p)\ )$ // Case 6
18. $list_p \leftarrow list_p.Add(Cell_i)$
19.   else
20. $If\ (\neg\ inside\ \&\ \neg left\ \&\ \neg down\ \&\ (X_i < X_p)\ )$ // Case 7
21. $list_p \leftarrow list_p.Add(Cell_i)$
22.   else
23. $list_p \leftarrow list_p.Add(Cell_i)$ // Case 8
24.   end
25.   End
26.   end
27.   end
28.   end
29.   end
30.   end

For cases where one or more *direct-connected neighbor* cells exist, algorithm finds the last *direct-connected neighbor cell* through X axe. Fig. 3-32.(b) shows that for $Cell_i$ in case the *right not directly-connected neighbor cells* are required, the algorithm finds the last *direct-connected neighbor cell* through X axe (incrementally), i.e $Cell_k$, and then calculates all *right not directly-connected neighbor cells* from $Cell_k$.

Similarly if for $Cell_k$ the *left not directly-connected neighbor cells* are required, the algorithm finds the last *direct-connected neighbor cell* through X axe (decrement), i.e $Cell_i$, and then calculates all *left not directly-connected neighbor cells* from $Cell_i$.

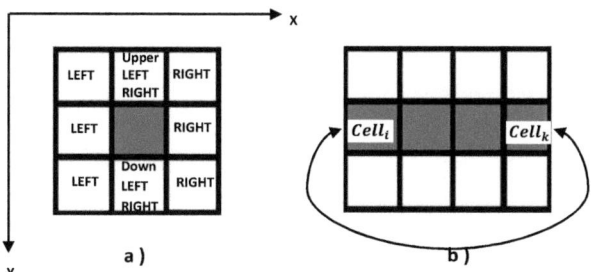

Fig. 3-32 Not directly-connected neighbour cells

Below at Alg. 3.4 is given a pseudo-code for finding the *last directed-connected* neighbor, as already explained above.

- The parameter P can be any calculated cell from Algorithm. 3.2.
- *left* is Boolean variable (true means the *left not directly-connected neighbor cells* are required, false means the right not directly-connected neighbor cells are required). (see explanation above)

Algorithm 3.4 : $Last\_\mathcal{D}_c\_Neighbor$ (P,left)

---
1. If (left)
2.   while ($\mathbb{N}(cell_i , P) \wedge F_i = true \wedge X_i = X_P + 1 \wedge Y_i = Y_P$)
3. $P_{new} \leftarrow cell_i$
4. $i \leftarrow i++$ // increment i
5.   end
6. else
7.   while ($\mathbb{N}(cell_i , P) \wedge F_i = true \wedge X_i = X_P - 1 \wedge Y_i = Y_P$)
8. $P_{new} \leftarrow cell_i$
9. $i \leftarrow i--$ // decrement i
10.   end
11. end

The sets containing the neighbor cells, obtained in this way are between-positions between SP and TP, and they are all connected to each other. The union of all these between-positions forms the path between COR position SP and the target position TP. The algorithm is very practicable at all *closed* objects having concave or convex polygons form. However in case there exist objects as open polygons and concave (Fig. 3-29 b. and c.), another issue is to find the "way-out" regions. Below it is described the heuristic method the algorithm uses in order to find such "way-out".

Suppose it is given a convex open object as in Fig. 3-33. Further let be $\Pi = \bigcup Cell_i \subset \mathcal{M}$ (see equation 3.12) set of all cells (in algorithm-structure defined as Point), so that for each $Cell_k$, $X_{ML} \leq X_i \leq X_{MR} \wedge Y_{MU} \leq Y_i \leq Y_{MD}$. ML, MR, MU and MD represent the most-left, most-right, most-upper and most-down cell of the object.

Let construct the number of interpolation lines between SP and any other point (cell) P_k , for $P_k \notin \Pi$. Generally the linear interpolation takes two points (in our case SP and P_k) and is given by :

$$y = y_{SP} + (y_k - y_{SP}) \frac{(x - x_{SP})}{(x_k - x_{SP})}, \text{ for every point in line } l_k \text{ (Fig. 3-33)}.$$

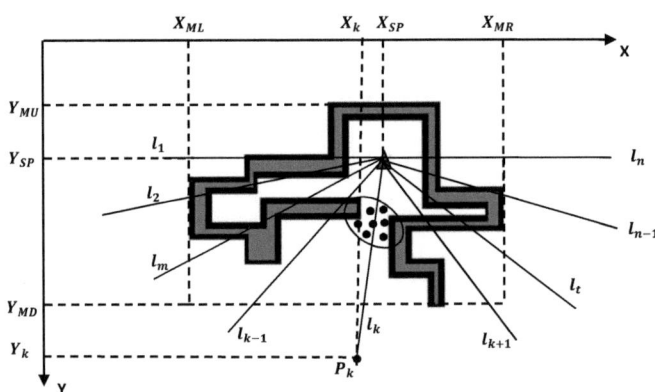

Fig. 3-33 The finding of "way-out" of an object

For the interpolated line l_k and the object, there are three different cases of intersection:
- If there is no intersection between line and boundary of object, then the whole region between SP and point P_k can be considered as "way-out". This case is trivial and very easy for calculation. There can be inserted also additionally some constraints in order to optimize and restrict found "way-out" region (see Alg. 3.5).
- If the number of intersections between interpolated line l_k and the boundary of the object, is even, then, as "way-out" can be considered all regions between each two pairs of intersection. To avoid the expansive handling calculation of such regions, in this work only the last (far way) "way-out" is considered as really "way-out" through line l_k.
- If the number of intersections between interpolated line l_k and the boundary of the object, is odd, then either the SP is inside the object, or point $P_k \in \Pi$. In both cases the algorithm should determine with error.

For the scanning of whole object for finding all "way-out", the number of several interpolated lines around the SP (360°) can be done. As "way-out" for the whole object is considered the "way-out" of the line with the smallest number of intersection. Below at Alg. 3.5 is given a pseudo-code for finding the *way-out region,* as already explained above. The parameter SP means Start Position, obj means Object, NR_k means the number of intersection between line Line$_k$ and obj means object. P is any cell (Point) outside the region Π (see above) over line Line$_k$ for a given angle.

Algorithm 3.5 : *Find_Way-Out (SP,obj)*

1. *angle ← start_value*
2. *Min ← start_min_value*
3. *while* $(angle < 360)$
4. $\text{Line}_k \leftarrow$ *Interpolation_Line*$(SP, P, angle)$ *such that* $P \notin \Pi$
5.   *if* $(NR_k \leftarrow \#$ *Intersection* $(\text{Line}_k, \text{obj}))$ *is odd then Error and Exit*
6.   *else*
7.   *If* $(NR_k \leftarrow$ *Intersection* $(\text{Line}_k, \text{obj}))$ *is 0 then* $\text{Line}_{res} \leftarrow \text{Line}_k$ *and Exit*
8.   *else*
9.   *If* $(NR_k < Min) \Rightarrow Min \leftarrow NR_k \wedge \text{Line}_{res} \leftarrow \text{Line}_k$
10.   *end*
11.   *end*
12.   *angle ← angle + value // Increment angle*
13. *end*
14. *//Restriction of the region*
15. *Previous_X ← Intersection* $(\text{Line}_{res-1}, \text{obj}).X \mid NR_{res} \neq NR_{res-1}$
16. *Previous_Y ← Intersection* $(\text{Line}_{res-1}, \text{obj}).Y \mid NR_{res} \neq NR_{res-1}$
17. *Next_X ← Intersection* $(\text{Line}_{res+1}, \text{obj}).X \mid NR_{res} \neq NR_{res+1}$
18. *Next_Y ← Intersection* $(\text{Line}_{res+1}, \text{obj}).Y \mid NR_{res} \neq NR_{res+1}$
19. *Way-out* $\leftarrow \bigcup_{i=1}^{N} cell_i \mid cell_i \in \mathcal{M} \wedge ($ *Previous_X* $< X_i <$ *Next_X* $)$
    $\wedge ($ *Previous_Y* $< Y_i <$ *Next_Y* $)$

Additionally, there is also the possibility to find any "way-out" near the desired region, especially in case the algorithm in this region cannot find further "between positions" between SP and TP . In those cases the hypertree decomposition will be applied.

And finally, there are cases where SP is in upper position then TP, i.e $y_{SP} > y_{TP}$. In such cases only the swapping of SP and TP positions is necessary, as the path remains unchanged.

Below at Alg. 3.6 is given a final pseudo-code for avoiding an object from COR.

The parameter obj means Object, start means Start-Position, target means Target-Position.

Algorithm 3.6 : *Object_Avoiding (obj, start, target,left)*

1. MR ← *Find_Most_Right_Cell of obj*
2. ML ← *Find_Most_Left_Cell of obj*
3. If (left)
4. $List_{down}$ ← **Most_Down(obj,ML,start,true)** //see Alg.3.1
5. $List_{upper}$ ← **Most_Upper(obj,ML,target,true)** //see Alg.3.2
6.   for ∀ $Cell_i$ ∈ $List_{down}$ ∪ $List_{upper}$
7.   P ← **Last_D_c_ Neighbor ($Cell_i$, true)** //see Alg.3.3
8. $List_{res}$ ← $List_{res}$. Add **(Not_D_c_Neighbor (P, true))** // see Alg.3.4
9.   end
10. else
11. $List_{down}$ ← **Most_Down(obj,ML,start,false)** //see Alg.3.1
12. $List_{upper}$ ← **Most_Upper(obj,ML,target,false)** //see Alg.3.2
13.   for ∀ $Cell_i$ ∈ $List_{down}$ ∪ $List_{upper}$
14.   P ← **Last_D_c_ Neighbor ($Cell_i$, false)** //see Alg.3.3
15. $List_{res}$ ← $List_{res}$. Add **(Not_D_c_Neighbor (P, false))** // see Alg.3.4
16.   end
17. end
18. $List_{way-out}$ ← **Find_Way_Out (start,obj)** //In case the Way-Out necessary. see Alg3.5
19. $List_{res}$ ← $List_{res}$ ∪ $List_{way-out}$

The avoiding algorithm for one object is already explained. Below it is described the whole algorithm from SP to TP, in case there exist more objects between them. In general the algorithm includes three steps:

First step: is the restricting of the searching cell-region. Especially finding of such paths where the COR due to the safety distance cannot pass. If the difference between some or all boundary cells of two different objects is smaller than COR size, objects can be merged into homogenous objects (Fig. 3-34 and chapter 3.2). In general the finding of homogenous objects can reduce the memory usage by the later path calculations. For the process of finding of the homogenous objects see chapter 3.2.

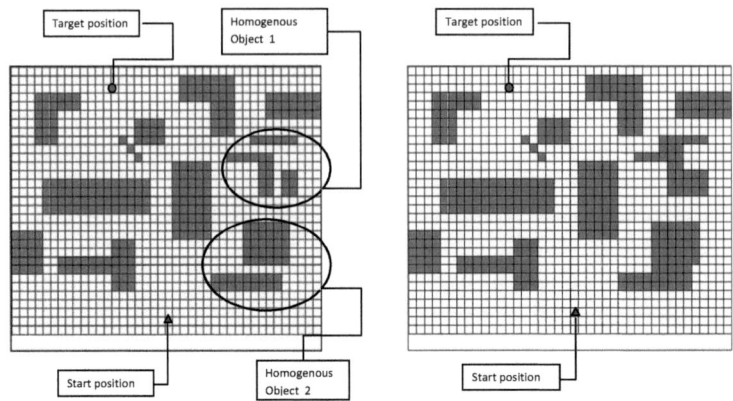

Fig. 3-34 The identification of the Homogenous Objects and the map after creation of Homogenous Objects.

Second step: is finding of several subsets $\mathcal{H}_i \subset \mathcal{M}$ for $i \in \{1,2,...,H\}$ and $H < N$ where $N = x_{max} * y_{max}$. One or more cells of each subset should be between-positions of the path. (Fig. 3-35).

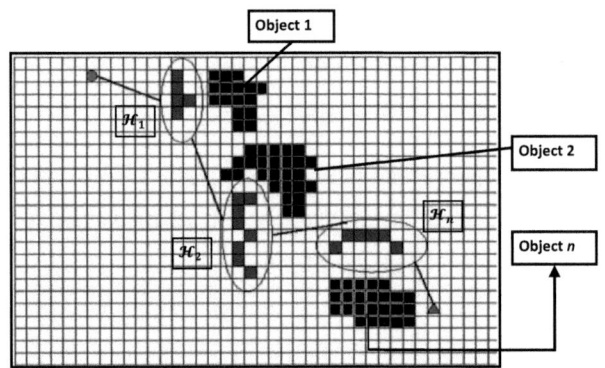

Fig. 3-35 Subsets as (between-positions) H_1 to H_n between Start and Target positions.

To find the subsets $\mathcal{H}_i$ as first the algorithm

1. Calculates the interpolation-line from Start position SP and Target position TP. (Fig. 3-36)

2. Calculates the intersection between this line and the first object occurred (near SP) on the map (Fig. 3-36).
3. On the found object the algorithm applies the avoiding algorithm for this object. At first object occurred, SP remain the same and TP became ML or MR depending on left or right avoiding. For the further avoiding of objects, recursively SP will substitute with TP from previous object.
4. Recursively apply point 1 to 4, having as new SP the calculated TP from point 3. The recursively approach exits in case the new TP became equally to the original TP

Below at Alg. 3.7 is given a pseudo-code for finding of the subsets $\mathcal{H}_i$ as described above.

The parameter SP means Start-Position around the object i.e. the Start-Position during the avoiding of only one object (see above point 3). TP means the absolute Target-Position i.e. the end-position where the COR should move.

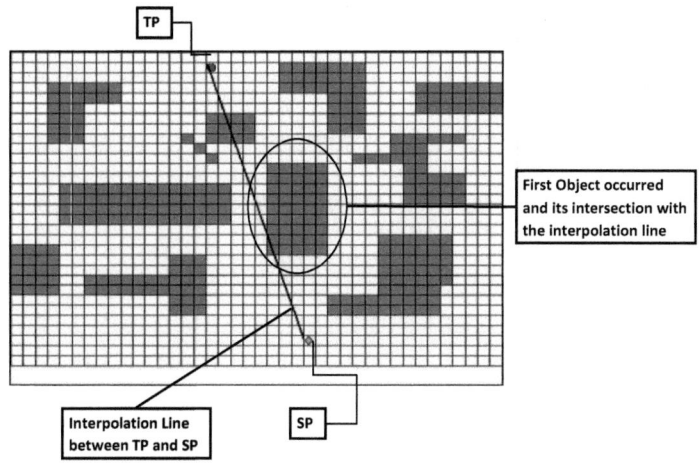

Fig. 3-36 Interpolation Line between SP and TP and its intersection with First Object occurred

Algorithm 3.7 : *Find_All_$\mathcal{H}_i$ (SP,TP)*

1. $Line_{interpol}$ ← InterpolLine (SP, TP) //Point 1
2. obj_{inter} ← *Find_Intersection* ($Line_{interpol,}$ $\mathcal{M}$) //Point 2
3. *left* ← *true*
4. *If* ($X_{obj_{inter}}$ $_{MR}$ < X_{TP} + X_{SP} − $X_{obj_{inter}}$ $_{ML}$) *left* ← *false*
5. *If* (left)
6. $list_{res}$ ← $list_{res}$ ∪ *Object_Avoiding* (obj_{inter} , SP, obj_{inter} ML, true)
   //see Alg. 3.6
7.   *If* (obj_{inter} ML == TP) EXIT Algorithm //see point 4
8.   *Find_All_$\mathcal{H}_i$* (obj_{inter} ML, TP) //call self recursively. See point 3 and 4.
9. *else*
10. $list_{res}$ ← $list_{res}$ ∪ *Object_Avoiding* (obj_{inter} , SP, obj_{inter} MR, false)
   //see Alg. 3.6
11.   *If* (obj_{inter} MR == TP) EXIT Algorithm //see point 4
12.   *Find_All_$\mathcal{H}_i$* (obj_{inter} MR, TP) //call self recursively. See point 3 and 4.
13. end

Third step: For all found subsets $\mathcal{H}_i \subset \mathcal{M}$, find the connections between each two neighbor subsets. In general the algorithm simply connects the two nearest elements from different subsets. But optionally there can be applied an extra optimality approach. This approach (for example Dijkstra algorithm) can be applied to find the shortest path only for the subsets elements (number of map-cells), and thus to win an extra optimality on this heuristic algorithm.

Finally, at Alg. 3.8 is given a finale pseudo-code for finding of the path between Start-Position SP and the Target-Position TP

Algorithm 3.8 : *Find_Path (SP,TP)*

1. $\text{List}_{\text{HomObjects}} \leftarrow$ Find_All_HomogenousObjects ( )   //see above First Step
2. $\text{List}_{\mathcal{H}} \leftarrow$ Find_All_$\mathcal{H}_i$ (SP,TP)   //see above Second Step and Alg. 3.7
3. $path \leftarrow leer$
4. If (option 1) // Simply connecting of cells
5.   for $\forall\, \mathcal{H}_k \in \text{List}_{\mathcal{H}}$
6.   for $\forall\, Cell_m ,\ Cell_n \in \mathcal{H}_k \wedge \mathbb{N}(Cell_m, Cell_n)$ //for each two neighbors cells
7. $path \leftarrow path \cup Connect\,(Cell_m, Cell_n)$
8.   end
9.   end
10.   for $\forall\, \mathcal{H}_i,\ \mathcal{H}_j \in \text{List}_{\mathcal{H}} \mid (\mathcal{H}_i$ and $\mathcal{H}_j$ neighbors)
11.   find $Cell_p \in \mathcal{H}_i \wedge Cell_q \in \mathcal{H}_j \mid$ Distance ($Cell_m, Cell_n$) is Minimal
12. $path \leftarrow path \cup Connect\,(Cell_p, Cell_q)$
13.   end
14. else   //optional : In case an optimal connection within list of between-position desirable // Applying of Dijkstra Algorithm over already found $\text{List}_{\mathcal{H}}$
15. $path \leftarrow$ Dijkstra_Algorithm ($\text{List}_{\mathcal{H}}$)   //see above Third Step

3.4.1.1 Summary

In this section a new heuristic algorithm for path planning is presented. This algorithm is based on geometrical intersection between start, target position and the obstacles between them. Is an obstacle between the start and target position, the algorithm try to find the free path between start and this obstacle. There are different heuristics applied: Is the target position on the right side in respect to the start position, than the algorithm find the most-right cell from the obstacle, otherwise the left-most cell will be found. The same procedure is done also for upper and down cells. Is the free path found, the algorithm continuous recursively for remain path between obstacle and target position.

The advantage of an approach is the runtime. Here only some area with cells on grid-map will be searched. Disadvantages are the restrictions by the very complex-formed areas and the memory efficiency, because for each obstacle must be stored cells, boundary cells, most left, most right, neighbor's cells etc.

3.4.2 A new heuristic algorithm for path planning based on hypertree decomposition [11]

In general the most of path planning algorithms produce a set of possible paths where using the graph representation the map-positions can be represented as vertices and paths can be represented as set of the edges. In case the optimal path is required, there can be applied a number of the typical search algorithms on these graphs, like depth-first-search, breadth-first search, Dijkstra algorithm , or even more sophisticate algorithms like A^*, D^*, Witkowski algorithm etc.

It is well known that the graphic structure of path planning problems can be a graph, hypergraph or in ideal case a tree structure. Although the graphs simplify problems, they are not always the best solution, otherwise the trees from the computational view are the best choice, but not always can represent a real problem. There are also a large number of real problems that are better represented by hypergraphs then by graphs (Dermaku et al.,2007), (Jeavons, Cohen and Gysens, 1991). One such case is representing of the grid-map (Fig. 3-28) by the hypergraph. A complete introduction about Hypergraphs and Hypertree decomposition is given in

---

[11] The algorithm is partially published at (A.Dermaku, 2013)

(Gottlob at. al, 1999). Below are given some basic definitions of hypergraph as well as its primal and dual form.

Definition3.8 (Jeavons., Cohen and Gysens, 1991) *A hypergraph H is an ordered pair (V,E) where V is a finite set of vertices and E is a set of edges, each of which is a subset of V.*

It can be differentiate between primal and dual graph of hypergraph. Graph *G(H)* is a primal (Geifman) graph of hypergraph $H = (V, E)$ if for any two vertices of the same hyperedge of hypergraph *H*, exists an edge in graph *G(H)* which connects these two vertices. (Harvey. and Ghose, 1999)

Formally , $G(H) = (V, E')$, $E' = \{\{X,Y\} | X,Y \in V, \exists e \in E : \{X,Y\} \in e\}$

Further it can be say that a graph *D(H)* is a dual graph of hypergraph *H* if its vertices are the edges of hypergraph *H* and for any two connected vertices in *D(H)* , exists at least one vertex of *H* in common. (Harvey and Ghose, 1999)

Formally, $D(H) = (V', E'')$ where $V' = E \wedge E'' = \{\{X,Y\} | X,Y \in E, \exists v \in V : v \in X \wedge v \in Y\}$

An example of hypergraph and its primal and dual graph can be found at Fig. 3-37 .

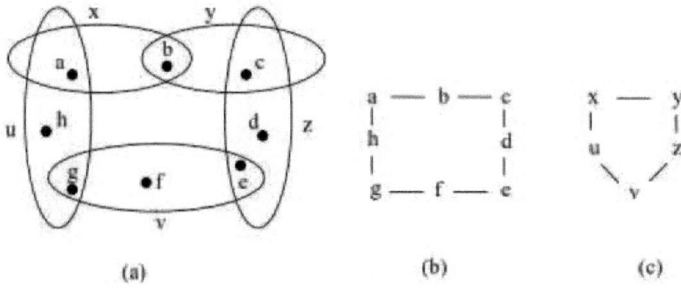

Fig. 3-37 A simple Hypergraph example (a), it's primal (b) and dual (c) graph

The hypergraphs primarily can be used for the CSP (Constraint Satisfaction Problems), because the structure of CSP problems can be much better represented by hypergraphs. It is well known that the higher number of cycles in hypergraphs (in general also in graphs) leads to increasing of the complexity. In the literature it can be found several decomposition methods which attempt to reduce the cycles in hypergraph, and if possible to generate an acyclic hypergraph.

For the relevance of this work, below will be introduced only the hypertree decomposition method. The hypertree decomposition method is developed by Gottlob et al. (Gottlob, Leone, Scarcello, 1999). According to the literature this method is the best decomposition method, which for a given constant k checks in polynomial time if the hypergraph has a hypertree width k. (Gottlob, Leone, Scarcello, 1999)

A *hypertree* for a hypergraph H is a triple $\langle T, \chi, \lambda \rangle$, where $T = (N, E)$ is a rooted tree, and χ *and* λ are labeling functions where $\lambda(p) \subseteq edges(H)$ *and* $\chi(p) \subseteq var(H)$. $Vertices\ (T)$ denotes the set of vertices N of T. Further for any $p \in N$, T_p denotes the subtree T rooted at p.

Definition 3.9 (Gottlob, Leone, Scarcello, 1999) *A hypertree decomposition of a hypergraph H is a hypertree* $HD = \langle T, \chi, \lambda \rangle$ *for H, which satisfies the following conditions.*

1. for each edge $h \in edges\ (H)$, there exists $p \in vertices\ (T)$ such that $var\ (h) \subseteq \chi(p)$. (we say p covers h);
2. for each variable $Y \in var\ (H)$, the set $\{p \in vertices\ (T) | Y \in \chi(p)\}$ induces a (connected) subtree of T;
3. for each $p \in vertices\ (T)$, $\chi(p) \subseteq var(\lambda(p))$;
4. for each each $p \in vertices\ (T)$, $var(\lambda(p)) \cap \chi(T_p) \subseteq \chi(p)$;

Below are given the explanation of the Definition-Conditions (Dermaku, 2007). The first condition for the new proposed algorithm in this work means that every cell (pixel) will be present at tree. Further, for every hyperedge the set of all its cells should be represented as such at least at one tree-vertex. The second condition (*connectedness condition*) means that all tree-vertexes are connected, i.e there exist at least two neighbor cells in two different tree levels (parent-child). This condition is the most relevant point for the new proposed approach, because it ensures the continuity of the path. Third condition means that every cell that occurs in tree vertex occurs also in one or several hyperedges which are part of this hypertree vertex. Fourth condition is not relevant for the new approach and can be ignored. In that case the decomposition is called *generalized hypertree decomposition*.

Definition 3.10 *Let be hypertree* $HD = \langle T, \chi, \lambda \rangle$ *and* $p_x, p_y \in T$ *any two vertices. In case that* p_x *and* p_y *are neighbors i.e.* p_x *is one level upper than* p_y, *then* p_x *is named the* father *and* p_y *its* child. Formally: $p_x = father(p_y)$ and $p_y = child(p_x)$.
In case they are not neighbors, then p_x and p_y are only in *Relation*. Formally: $p_x \, R \, p_y$.

In Fig. 3-38 is given an example for (a) Hypergraph and (b) its Hypertree Decomposition of width 2. It can be very easy shown that:
- First condition is fulfilled. Every variable $X_1, X_2, \ldots, X_{10} \in$ $Hypergraph$ is present at Hypertree, and for every hyperedge its variables are represented at least at one tree vertex. For example $S_1 = \{X_1, X_2, X_3\}$ can be found at vertex p_4 and p_3, or S_6 and S_7 with all their respective variables can be found at vertex p_1, etc.
- Second Condition is also fulfilled. The root of Hypertree (p_1) is connected with its left child (p_2) vertex because

S_6 is connected with S_9 and S_7 is connected with S_8. In general only one connectedness is enough. The root of Hypertree is connected with its right child (p_3) vertex because S_3 is connected with S_6 and S_4 is connected with S_7. Finally the second level of hypertree (p_3) is connected with the leaf of hypertree p_4, because S_1 is connected with S_3 and S_1 is connected with S_4. (Fig. 3-38 (b))

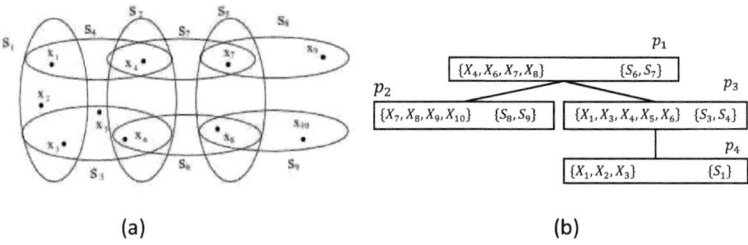

Fig. 3-38 A Hypergraph example (a) and its Hypertree Decomposition of width 2 (b)

- Third condition is also fulfilled. Each variable $X_i \in Hypertree$ occurred at least at one Hypergraph-Edge (see Fig. 3-38 (a))
- Fourth condition is not relevant for this work.

In this example (Fig. 3-38) the hypertree vertexes p_1, p_2, p_3 are covered in worst case with two hypertree edges i.e $\{S_6, S_7\}$ or $\{S_8, S_9\}$ or $\{S_3, S_4\}$ so the hypertree-width is 2. The smaller hypertree-width leads to efficiently solutions and smaller runtime.

The problem of Hypertree Decomposition is the runtime of its exact decomposition algorithm. Indeed, the exact decomposition algorithm *opt-k-decomp* developed from Gottlob (Gottlob, Leone, Scarcello, 1999), runs in $O(m^{2k}v^2)$ time, where m and v are the number of edges and number of vertices of Hypergraph, respectively. It can be seen that although the *opt-k-decomp* runs in polynomial time, the parameter *k* (desired hypertree width) remains at the exponent of runtime, what make the *opt-k-decomp* not practicable for the proposed algorithm in this work. To overcome the problem of runtime and to make the method useful also for large problems, the DBAI research group (see Dermaku et al.,2007)[12] developed several heuristic approaches. These approaches are the best existing heuristic methods in that field. In general they are influenced by the Hypergraph structure but in some cases achieve optimal decomposition results, and thus in polynomial time. The heuristics, especially two of them, Hypertree Decomposition based on HMETIS partitioning, and Hypertree Decomposition based on Bucket Elimination have shown the best results from heuristics.

The implementation of the heuristics from DBAI research group (see Dermaku at.al.,2007), is used in this work to implement the proposed new heuristic algorithm for path planning.

Below in this work, it is shown the process of using of the hypergraph structure for the grid-map as well as its hypertree decomposition. Further there are given the advantages and eventual disadvantages of the new heuristic algorithm proposed in this work comparing to other existing algorithms.

Let be *occupancy grid* map the same as proposed in section 3.2 and the same as used also in section 3.4.1. Basically the environment is divided into cells (pixels) of equal size $cell_i = \{x_i, y_i, F_i\}$ (Fig. 3-7 in chapter 3.2).

Similarly to section 3.4.1, the used grid map used for this approach can be found at (Fig. 3-39). The map here is also represented as a set of the elements (cells) :

$\mathcal{M} = \bigcup_{i=1}^{N} cell_i$, $i \in \{1,2, ..., N\}$ (see equation 3.12 in section 3.4.1)

(3.14)

---

[12] The author of this work was member of the DBAI research team

in pixel coordinate system where $N = x_{max} * y_{max}$. Values x_{max}, y_{max} denote the maximal coordinate values in x and y axe.

For $cell_i$ the relation between cell-index i and cell-coordinate values x_i, y_i can be calculated from equations $x_i = i \bmod x_{max}$ and $y_i = i / x_{max}$, where, $i = y_i * x_{max} + x_i$ (see also chapter 3.2)

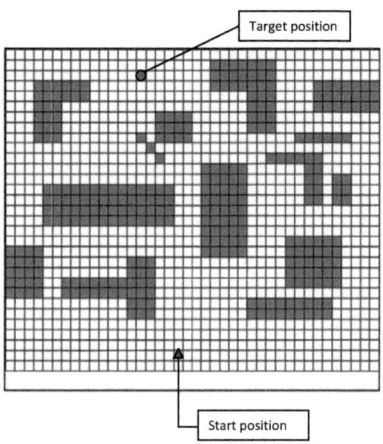

Fig. 3-39 Binary Occupancy grid map (see chapter 3.2)

Also at this approach, similarly to section 3.4.1, the COR is represented by a triangle mask in the greed map within a greed cell, and its size is assumed to be equally to the size of a cell.

As first it is described the algorithm for hypergraph presentation of the grid-map. Let be Γ the union of all not filled cells from grid-map. Formally:

$$\Gamma = \bigcup_{i=1}^{N} cell_i \mid cell_i = \{x_i, y_i, F_i\} \wedge F_i = False \wedge cell_i \in \mathcal{M}. \quad (3.15)$$

From Equations (3.14 and 3.15) follows $\Gamma \subset \mathcal{M}$. The new heuristic algorithm consists from two main steps. First step is creating of hypertree from hypergraph using hypertree decomposition based on heuristics. Second step is finding of paths within every hypertree-vertex. To find the optimal path between two hypertree vertexes also a special approach can be applied (Dijktstra as exact approach or any another heuristics)

First step (creating of hypertree from hypergraph) has four sub-steps. In the sub-steps one to three, from map-grid will be created Hypergraph, and in fourth sub-step from Hypergraph using hypertree decomposition method will be created Hypertree (see Definitions 3.8 and 3.9).

Sub-step 1 is the numbering of all cells $cell_i \in \Gamma$. The numbering begins from left-up to down-right on the grid-map.
Formally:
$\forall \ cell_i \in \Gamma, \ i \in \{1,2,...,N\}, \ cell_i \leftarrow Nr_i \mid Nr_{i+1} - Nr_i = 1$
$cell_i \leftarrow Nr_i$ means $cell_i$ is denoted with number equally to Nr_i .

Sub-step 2 is the dividing of the grid-map into several areas, where each area represents a hyperedge and the empty cells within this area are the vertices of the respective hyperedge.
Formally: $\Gamma = \bigcup_{i=1}^{p} E_i$ where $E_i = \bigcup_{k=1}^{N} cell_k \mid cell_k \in \Gamma$

Sub-step 3 is process of creating of Hypergraph . The areas found in second step are the hyperedges and as vertices remains all cells from Γ . Beginning from up-down and from left-right the current hyperedge includes also the neighbor cells (vertices) from neighbor hyperedge. These so-called common vertices, ensures the possible connectivity.
Formally: $H = (V, E) \ where \ V = \Gamma \land E = \bigcup_{i=1}^{n} E_i$

Sub-step 4 is process of creating of Hypertree from Hypergraph . In this work, as already mentioned, in order to generate the Hypertree it is used the Hypertree Decomposition method. To overcome the runtime problem, the heuristics algorithms based on HMETIS partitioning and Bucket Elimination are used. The Hypertree must fulfill all conditions from Definition. 3.9.

Below at Alg. 3.9 is given a pseudo-code for creating of hypertree from a hypergraph.

Input parameter $\mathcal{M}$ means Grid-Map.

Algorithm 3.9 : Create_Hypertree ($\mathcal{M}$)

---
1. $\text{List}_{\text{FreeCells}} \leftarrow$ leer $\wedge$ $i \leftarrow 1$ //Initialize a List of not filled cells, and index i
2. for $\forall\ Cell_i \in \mathcal{M} \wedge F_i = False$
3. $\quad\quad\text{List}_{\text{FreeCells}} \leftarrow \text{List}_{\text{FreeCells}} . \text{Add}\ (Cell_i)$ //see above Equation 3.15
4. $\quad\quad i \leftarrow i++$ //increment i
5. end
6. $Hypergraph\_H \leftarrow$ **Create_Hypergraph (** $\text{List}_{\text{FreeCells}}$ **)** //see Alg. 3.10
7. //Applying of Heuristic approaches HypertreeDecomposition based on HMetis and //BucketElimination
$Hypertree\_HG \leftarrow HypertreeDecomposition\ (Hypergraph_H)$
8. $return\ Hypertree\_HG$

Below at Alg. 3.10 is given a pseudo-code for creating of hypeedges hypergraph. Input parameter $\text{List}_{\text{FreeCells}}$ means List of all free cells in Grid-Map.

Algorithm 3.10 : *Create_Hypergraph (* List$_{FreeCells}$ *)*

1. List$_{HypEdges}$ ← leer   //*Initialize a List of of not filled cells*
2. *for* $i \leftarrow 1$ *to* $i \leftarrow NrHypEdges$ *do*   //*Create all hyperedges. NrHypEdges*
   //*represents the number of dividing area of map (see substep 2 and 3)*
3.   *for* $\forall\ Cell_k\ \in\ $List$_{FreeCells} \wedge\ Cell_k\ \in\ List[i]$ *do*
4. $List[i] \leftarrow List[i].Add(Cell_k)$
5.   *end*
6.   List$_{HypEdges}$ ← List$_{HypEdges}.Add(List[i])$
7. *end*
8. List$_{HypEdges}$ ← leer   //*Initialize a List of of not filled cells*
9. *for* $i \leftarrow 1$ *to* $i \leftarrow NrHypEdges$ *do*   //*Create all hyperedges. NrHypEdges*
   //*represents the number of dividing area of map (see substep 2 and 3)*
10.   *for* $\forall\ Cell_k\ \in\ $List$_{FreeCells} \wedge\ Cell_k\ \in\ List[i]$ *do*
11. $List[i] \leftarrow List[i].Add(Cell_k)$
12.   *end*
13.   List$_{HypEdges}$ ← List$_{HypEdges}.Add(List[i])$
14. *end*
15. *for* $\forall\ List_i\ \in\ $List$_{HypEdges}$ *do*   //*Including of the neighbor cells (see substep 3)*
16.   *for* $\forall\ List_j\ \in\ $List$_{HypEdges}$ *do*
17.   List$_{inter}$ ← $(variables(List_i)\ \cap\ variables\ (List_j\ )) \neq \emptyset\ \wedge (i > j)$ *if*
18. $List_i \leftarrow List_i\ .Add($List$_{inter})$
19.   *end*
20.   *end*
21. *end*
22. *return* List$_{HypEdges}$

To make more understandable the process of creating of hypergraph , let suppose the grid map as in Fig. 3-40. In the *first sub-step* all empty cells are described with its numbers beginning from left-up to down-right, in our case from 1 to 218. In *second sub-step* the grid-map is divided into several areas. Note that the qualitative dividing of map into areas in general impacts the optimality of the hypertree decomposition. For simplicity in our case it is used only dividing into 6 areas, but the higher number could impact the heuristic algorithm. As result there are 6 hyperedges E_1 to E_6 (Fig. 3-40) where E_x corrosponds to $X - area$

In *third sub-step* it is created Hyeprgraph $H = (V, E)$, $V = Cell_i$ for $i \in \{1, 2 \ldots 218\}$ and $E = E_1 \cup E_2 \cup \ldots E_6$. Below are given 6 Hyperedges as well as theirs Vertices (cells) and Common Cells between neighbor Hyperedges.

$E_1 = \{\ 1, 2, \ldots\ 9, 20, \ldots 23, \ldots, 86, 87\ \}$, $E_2 = \{\ 10, \ldots, 14, \ldots, 84, \ldots, 92\}$,
$E_3 = \{\ 15, \ldots, 19, \ldots, 93, \ldots, 97\}$, $E_4 = \{\ 98, \ldots, 106, \ldots, 200, \ldots, 208\}$,

$E_5 = \{\ 107, \ldots, 112, \ldots, 209, \ldots, 214\}$, $E_6 = \{\ 113, \ldots, 116, \ldots, 215, \ldots, 218\}$

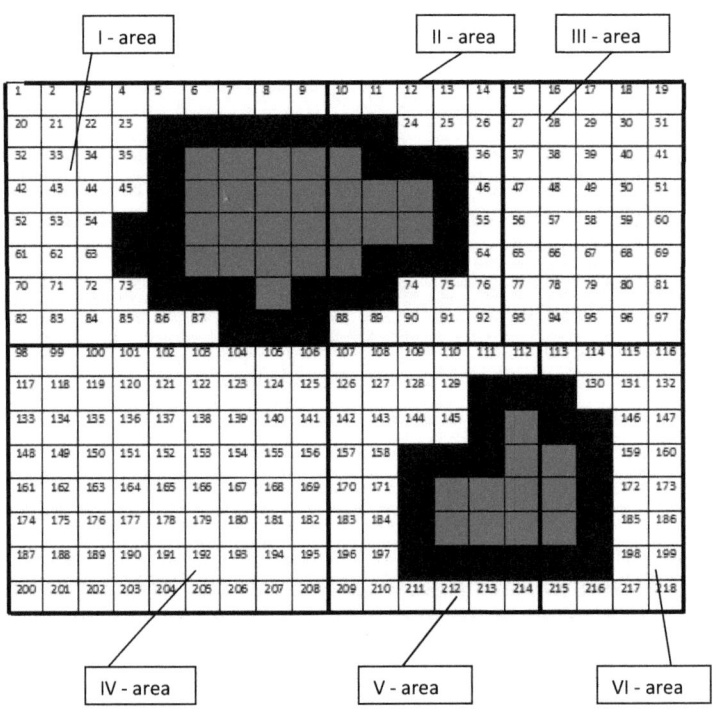

Fig. 3-40 Numbering of the empty cells of the grid-map

The common cell between E_1 and E_2 will be considered the cell with number 10. As the left hyperedge is the Hyperedge E_1, so this hyperedge will be extended with common cells, $E_1 = E_1 \cup \{10\}$.

The common cells between E_1 and E_4 will be considered the cells with number 98,99,100,101,102,103. The upper hyperedge is the Hyperedge E_1, so this hyperedge will be extended with common cells, $E_1 = E_1 \cup \{98,99,100,101,102,103\}$.

The process of extensions with common vertices will be done for all hyperedges. Once this process is finished, the final Hypergraph with Hyperedges and all its Vertices is created. In *fourth sub-step* it is created Hyepertree from Hypergraph, using heuristic approaches like HMetis and Bucket Elimination. As already mentioned, these heuristics are detailed introduced in chapter 3, so here will be no further explanation. From Fig. 3-42 it can be proved that all Hypertree condition are fulfilled, and thus the first step of the new proposed heuristic algorithm is finished. Further the second step of the algorithm can begin.

Second step is finding of paths within every hypertree-vertex between vertex containing start position (start cell) and vertex containing target position (target cell). The second step consists from two sub-steps.

> ***Sub-step 1*** is process of searching of *start cell* $\equiv cell_x \leftarrow Nr_x$ and *target cell* $\equiv cell_y \leftarrow Nr_y$ within hypertree. For this purpose the new proposed algorithm uses DFS (Depth First Search) approach on the hypertree. In case there are more hypertree vertices that contain start and target positions, the vertices will higher level will be marked. Finally the path between them should be finding. Let be HD the created Hypertree where $p_x, p_y \in vertices\ (HD)$ and $cell_x \in p_x$, $cell_y \in p_y$. In case that start and target cells $cell_x$, $cell_y$ are in different tree vertices, the vertices p_x and p_y can be either father-

child or only in relation (see Definition 3.10). First case is trivial, i.e the algorithm is restricted only on these two vertices p_x, p_y. In the second case, where $p_x R p_y$ the algorithm finds all vertices between p_x, p_y.

Formally: $path_{x,y} = \bigcup_x^y p_i \mid \forall p_m, p_n \in path_{x,y} \Rightarrow$
$p_m = father(p_n)$ or $p_m = child(p_n)$ or
$\exists p_k \in path_{x,y} \mid path_{x,y} = path_{x,k} \cup path_{k,y}$

Let suppose :
$path_{x,y} =$
$\{p_x, p_{x+1}, p_{x+2}, \ldots, p_{k-2}, p_{k-1}, p_k, p_{k+1}, p_{k+2}, \ldots, p_{y-2}, p_{y-1}, p_y\}$
, where p_k is root of the hypertree, i.e the p_k has a level 0. Then for $path_{x,y}$ the following conditions must be fulfilled:

1. $p_k = father(p_{k-1})$, $p_{k-1} = father(p_{k-2})$ ... $p_{x+2} = father(p_{x+1})$, and $p_{x+1} = father(p_x)$
2. $p_k = father(p_{k+1})$, $p_{k+1} = father(p_{k+2})$ ... $p_{y-2} = father(p_{y-1})$, and $p_{y-1} = father(p_y)$

Sub-step 2   In each vertex of the hypertree, found in the *Sub-step1*, the algorithm finds the shortest path i.e. set of variables (cells). Let be Hypertree $HD = \langle T, \chi, \lambda \rangle$ and let be $cell_x \leftarrow Nr_x$ start-position in hypertree vertex p_x and $cell_y \leftarrow Nr_y$ target-position in hypertree vertex p_y. Follows $cell_x \in \chi(p_x)$ and $cell_y \in \chi(p_y)$.

In general the algorithm in *Sub-step 2* applies four steps.

1. If $cell_x \leftarrow Nr_x$ has a lower level in hypertree then $cell_y \leftarrow Nr_y$ the algorithm creates a vertex-path $path_{x,y} = \{p_x, p_{x+1}, p_{x+2}, \ldots, p_k, p_{k+1}, \ldots p_y\}$ otherwise $path_{x,y} = \{p_y, p_{y+1}, p_{y+2}, \ldots, p_k, p_{k+1}, \ldots p_x\}$, (Fig. 3-41).

For each two successive hypertree vertices p_k, p_{k+1} follows either $p_k = father(p_{k+1})$ or $p_k = child(p_{k+1})$. Indeed for algorithm is not important if path is calculated from start to target position or from target to start position, because the path remains the same. However due to the calculation rules, it is decided to start from cell with smaller hypertree level value to the cell with larger level value.

2. Is $path_{x,y} = \{p_x\}$, i.e the start and target positions are within same hypertre vertex, algorithm applies at current vertex p_x, directly step 3.1 where $cell_q \equiv cell_y$ otherwise :

The algorithm finds the Common Cells $CS = \chi(p_x) \cap \chi(p_{x+1})$ Recursively while $CS \neq \emptyset$ repeat $CS = CS \cap \chi(p_{x+i})$ and increment i. Finally finds the last hypertree vertex $p_k \in path$ where $\neq \emptyset$. Then, at current vertex p_x will be applied step 3 .

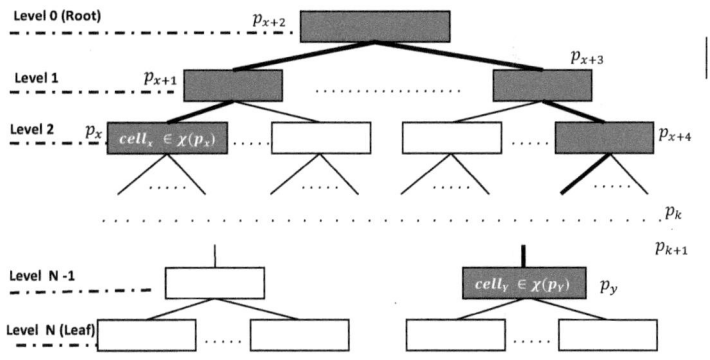

Fig. 3-41 Example of a Hypertree Decomposition with its Levels and a possible Path

3. As first algorithm tries to find $cell_q \in CS$ where $cell_q \leftarrow Nr_q$, such :

(i) Nr_q is between (Nr_x and Nr_y) $\wedge$ $|Nr_y - Nr_q| \leftarrow$ Minimal, or

If cell not found then tries to find $cell_q \in CS$ for:

(ii) $Nr_q > Nr_x \wedge |Nr_q - Nr_y| \leftarrow$ Minimal, or

If cell not found at 3.2.2 then tries to find $cell_q \in CS$ for:

(iii) $Nr_q < Nr_x \wedge |Nr_y - Nr_q| \leftarrow$ Minimal

3.1 Is $p_k \in path$ where $CS \neq \emptyset$ (from step 2) then algorithm finds path between $cell_x$ and $cell_q$ within the hypertree vertex p_x. There is not necessary to calculate all possible cells, but only the neighbor cells. Formally:

(1) Find the cell $cell_k \in \chi(p_x) | \mathbb{N}(cell_x, cell_k)$ and $|Nr_q - Nr_k| \leftarrow$ Minimal. If $cell_k \equiv cell_q$ break the process and return path, otherwise

(2) While $cell_k \not\equiv cell_q$ repeat point (1) for new $cell_s | \mathbb{N}(cell_k, cell_s)$.

(3) If $cell_s \notin \chi(p_x)$ or $cell_s \equiv cell_k$ (the process is in cycle), then go back to the last cell $cell_k$ and repeat (1) for another $cell_r | \mathbb{N}(cell_k, cell_r) \wedge cell_r \not\equiv cell_s$. This approach is very similar to backtracking algorithm.

(4) If there is no cell that fulfills conditions 3.1.(1) to 3.1.(3) then go back to the last found $cell_k \in CS$ and move to p_{x+1}. Continue the process beginning from 3.1.(1) for the new hypertree vertex p_{x+1}

4. If $cell_q$ (obtained from step 3) and $cell_y$ are in different vertices, then repeat the process recursively from step 2, but now for start position $cell_q$ and target position $cell_y$,

otherwise find path as described in point 3.1 between $cell_q$ and $cell_y$.

Although the algorithm is based on heuristic approaches, and only the neighbors as in point 3 and 4 described, are used, the paths are suboptimal, and the differences from optimal one are very small. Sometimes can be achieved also an optimal solution.

In general, in case that an optimal solution is required, then at step 3 and 4 not only some neighbors should be taken in consideration, but all of them, what increases the runtime. But also in such cases, there are much less calculations necessary then for example exact Dijkstra Algorithm, because the number of cells contained by the path-hypertree in ideal case is much smaller then total number of map-cells . And finally, one of the main property of the Hypertree Decomposition with relevance for the algorithm in this work is the connectivity between vertices, which ensures that always exist a path between two cells if they are within vertex path obtained from step 1.

To make more understandable the process of finding of paths in hypertree as in *Second step* is already described, below can be found the example of path finding between cell with number 58 (start cell) and cell with number 53 (target cell) (see Fig. 3-40) .

The hypertree decomposition is given in Fig. 3-42.

From *sub-step1* after Hypertree decomposition follows: *start position*≡ $cell_{58}$, *target position*≡ $cell_{53}$, *start vertex* ≡ p_2, *target vertex* ≡ p_5.

From *sub-step2* follows:

1. $path_{58,53} = \{p_2, p_1, p_3, p_5\}$, because the p_2 has a lower level then p_5. The algorithm should find the path between cell with number 58 in hypertree vertex p_2 and cell with number 53 in hypertree vertex p_5 .
2. $Common\ Cells\ CS = \chi(p_2) \cap \chi(p_1) \cap \chi(p_3) = \{112, 113, 215\}$.
3. i) There is no $cell_q \in CS$ between 58 *and* 53.

As there is no $cell_q$ which fulfilled the condition under i) algorithm tries with ii)

ii)
$58 < Nr_q \wedge (Nr_q - 53) \leftarrow Minimal \Rightarrow Nr_q = 112 \; and \; cell_q \equiv cell_{112}$

The cell $cell_{112} \in \chi(p_3)$, therefore the last hypertree vertex containing the $cell_{112}$ is p_3.

3.1.(1) Neighbor $cell_k$ of $cell_{58}$ in vertex p_2 where $|112 - Nr_k| \leftarrow Minimal$, is $cell_{67}$.

(2) $cell_{67} \not\equiv cell_{112}$, repeat 3.1.(1) and find new $cell_{79} \not\equiv cell_{112}$ then $cell_{79} \not\equiv cell_{112}$ then $cell_{95} \not\equiv cell_{112}$ then $cell_{114} \not\equiv cell_{112}$ then $cell_{113} \not\equiv cell_{112}$ and finally $cell_{112} \equiv cell_{112}$. The found path from 3.1 is :

$path = (cell_{58}, cell_{67}, cell_{79}, cell_{95}, cell_{114}, cell_{113}, cell_{112})$

4. $cell_q \leftarrow 112$ is in vertex p_3 (obtained from step 3) and $cell_y \leftarrow 53$ is in vertex p_5. They are in different vertices, therefore the process will be repeated recursively from point 2, for start position with $cell_{112}$ and target position with cell $cell_{53}$.

Repeat at:

2. *Common Cells* $CS = \chi(p_3) \cap \chi(p_5) = \{10, 107, 126, 142, 157, 170, 183, 196, 209\}$.

3. i) There is $cell_{107} \in CS$ between $112 \; and \; 53$, therefore $cell_q \equiv cell_{107}$

3.1. (1). The Neighbor $cell_k$ of $cell_{112}$ in vertex p_3 where $|107 - Nr_k| \leftarrow Minimal$, is $cell_{111}$.

(2). Because $cell_{111} \not\equiv cell_{107}$ repeat 3.1.(1) and find new $cell_{110} \not\equiv cell_{107}$ then $cell_{109} \not\equiv cell_{107}$ then $cell_{108} \not\equiv cell_{107}$ and finally $cell_{107} \equiv cell_{107}$.

Because $cell_{107} \equiv cell_q \leftarrow 107$, break the process and moves to p_5.

The found path from 3.1.2 is :
$path = (cell_{111}, cell_{110}, cell_{109}, cell_{108}, cell_{107})$

4. $cell_q \leftarrow 107$ is in vertex p_5 (obtained from point 3) and $cell_y \leftarrow 53$ is also in vertex p_5, therefore the condition 3.1 will be applied.

3.1 (1) The Neighbor $cell_k$ of $cell_{107}$ in vertex p_5 where $|53 - Nr_k| \leftarrow Minimal$, is $cell_{106}$.

3.2 (2) Because $cell_{106} \not\equiv cell_{53}$, repeat 3.1.(1) and find new cell $cell_{105} \not\equiv cell_{53}$, then $cell_{104} \not\equiv cell_{53}$, then $cell_{103} \not\equiv cell_{53}$, then $cell_{87} \not\equiv cell_{53}$, then $cell_{86} \not\equiv cell_{53}$, then $cell_{85} \not\equiv cell_{53}$, then $cell_{73} \not\equiv cell_{53}$, then $cell_{72} \not\equiv cell_{53}$, then $cell_{63} \not\equiv cell_{53}$, then $cell_{54} \not\equiv cell_{53}$, and finally $cell_k = cell_{53} \equiv cell_{53}$.

The found path from point 4 is :

$path = (cell_{106}, cell_{105}, cell_{104}, cell_{103}, cell_{87}, cell_{86},$
$, cell_{73}, cell_{72}, cell_{63}, cell_{54}, cell_{53})$.

And finally the algorithm gives back the path:

$path$
$= (cell_{58}, cell_{67}, cell_{79}, cell_{95}, cell_{114}, cell_{113}, cell_{112}, cell_{111}, cell_{110},$
$cell_{109}, cell_{108}, cell_{107}, cell_{106}, cell_{105}, cell_{104}, cell_{103}, cell_{87}, cell_{86},$
$cell_{85}, cell_{73}, cell_{72}, cell_{63}, cell_{54}, cell_{53})$.

The found path is also the optimal one. (Fig. 3-43).

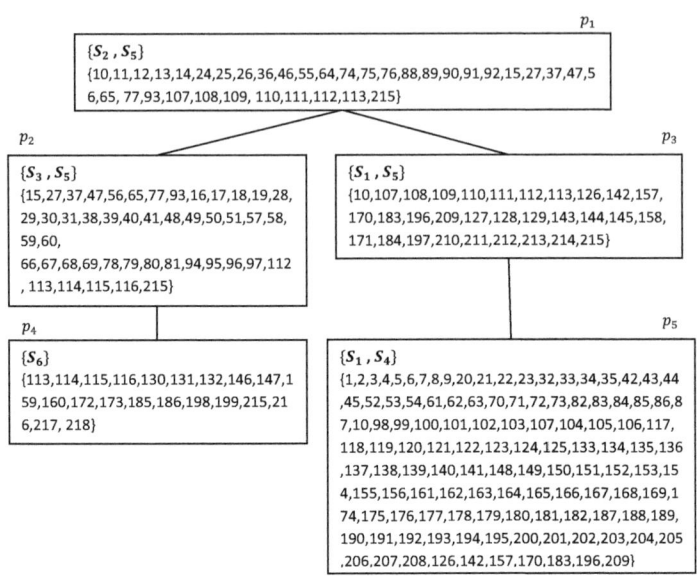

p_1

$\{S_2, S_5\}$
{10,11,12,13,14,24,25,26,36,46,55,64,74,75,76,88,89,90,91,92,15,27,37,47,5
6,65, 77,93,107,108,109, 110,111,112,113,215}

p_2 p_3

$\{S_3, S_5\}$
{15,27,37,47,56,65,77,93,16,17,18,19,28,
29,30,31,38,39,40,41,48,49,50,51,57,58,
59,60,
66,67,68,69,78,79,80,81,94,95,96,97,112
, 113,114,115,116,215}

$\{S_1, S_5\}$
{10,107,108,109,110,111,112,113,126,142,157,
170,183,196,209,127,128,129,143,144,145,158,
171,184,197,210,211,212,213,214,215}

p_4 p_5

$\{S_6\}$
{113,114,115,116,130,131,132,146,147,1
59,160,172,173,185,186,198,199,215,21
6,217, 218}

$\{S_1, S_4\}$
{1,2,3,4,5,6,7,8,9,20,21,22,23,32,33,34,35,42,43,44
,45,52,53,54,61,62,63,70,71,72,73,82,83,84,85,86,8
7,10,98,99,100,101,102,103,107,104,105,106,117,
118,119,120,121,122,123,124,125,133,134,135,136
,137,138,139,140,141,148,149,150,151,152,153,15
4,155,156,161,162,163,164,165,166,167,168,169,1
74,175,176,177,178,179,180,181,182,187,188,189,
190,191,192,193,194,195,200,201,202,203,204,205
,206,207,208,126,142,157,170,183,196,209}

Fig. 3-42 Hypertree Decomposition for Hypergraph given in Fig.3-42

Fig. 3-43 Path calculated from Algorithm based on Hypertree Decomposition for cells between number 53 and 58

The path in Fig. 3-43, although it is obtained from a heuristic algorithm, is an optimal path.

3.4.2.1 Summary

In this section a new heuristic algorithm for path planning is presented. This algorithm is based on hypertree decomposition. Hypertree decomposition is a method which decomposes a hypergraph into hypertree. The main advantage of such method is the runtime. The solution of the problems formulated as hypertree has significantly better runtime than problems formulated as graphs. The main disadvantage is the time needed for the process of decomposition. For this reason a several heuristic decomposition algorithms are developed. Once the decomposition of hypergraph into hypertree is done, all further calculation can be done on hypertree and not in graph. In case of multi agent system robots, where more robots moves on the same area these approaches can be ideal solution because of the improving of time-calculation.

Chapter 4

4 Hardware Implementation

In this chapter a mobile robot as a prototype of a demining robot named HUMI, as well as a Humanoid Robot Archie will be introduced. Both these robots are designed and implemented at the IHRT (Institute at Handling Robot and Technology), Vienna University of Technology.

The whole description of design and implementation of the robots is already from several authors introduced, so in this work, the main focus on HUMI and Archie is theirs kinematic and dynamics. The aim of the work in this chapter is finding and describing of moving model of robots in respect to calculated path on map, as already described in previous chapter.

In order to reduce the errors occurred during the following of trajectory by the robot (distance error, angle error, etc), on this work a new implementation of Fuzzy Logic controller for HUMI is done. This approach shows a significant improving of moving correctness for larger velocities of mobile robots, i.e. for HUMI velocity greater than 10 m/s error values are significantly reduced, otherwise for the smaller velocities the differences are relative small.

For the same reason a PID (Proportional – Integral – Derivative Controller) for robot Archie is implemented. This generic control loop feedback mechanism calculates the error values of hip- angle, knee-angle, and ankle-angle and attempts to minimize the errors by adjusting the process control inputs from ELMO motion control.

4.1 HUMI – A Demining Robot

IHRT institute (Vienna University of Technology) has developed a prototype of a demining robot (Silberbauer, 2008). This prototype consists of a platform and a metal detection sensor and is equipped with an internal micro controller as well as internal sonar sensors, position speed encoders and a battery pack for network-independent and autonomous operation.
An introduction for HUMI (Fig. 4-1) is given in chapter 2.

Fig. 4-1 HUMI

4.1.1 Analysis of HUMI-s kinematic model

Let be the linear velocity of the HUMI at point B with coordinates x_B, y_B (Fig. 4-2), $v_B = v$ and steering angle of front wheels: ϕ respectively $\dot{\phi}$.
From the non–holonomic constraints follows that the HUMI move in the direction normal to the axis of the driving wheels (rear axle) and it is assumed that no-slipping occurred.

From the condition of no-slipping towards the rear wheel axis CD at point B (Fig. 4-2) follows : [13]

$$\dot{x}_B \cdot \sin(\theta) - \dot{y}_B \cdot \cos(\theta) = 0 \tag{4.1}$$

From the condition of no-slipping towards the front wheel axis EF at point A (Fig. 4-2) follows

$$\dot{x}_A \cdot \sin(\theta + \phi) - \dot{y}_A \cdot \cos(\theta + \phi) = 0 \tag{4.2}$$

Further follows:

$$x_A = x_B + 2b \cdot \cos(\theta), \quad y_A = y_B + 2b \cdot \sin(\theta) \; . \tag{4.3}$$

$$\dot{x}_A = \dot{x}_B - 2b \cdot \dot{\theta} \cdot \sin(\theta), \quad \dot{y}_A = \dot{y}_B + 2b \cdot \dot{\theta} \cdot \cos(\theta) \tag{4.4}$$

$$\dot{x}_B \cdot \sin(\theta+\phi) - 2b \cdot \dot{\theta} \cdot \sin(\theta) \cdot \sin(\theta+\phi) - \dot{y}_B \cdot \cos(\theta+\phi) - 2b \cdot \dot{\theta} \cdot \cos(\theta) \cdot \cos(\theta+\phi) = 0$$

$$\dot{x}_B \cdot \sin(\theta+\phi) - \dot{y}_B \cdot \cos(\theta+\phi) - 2b \cdot \dot{\theta} \cdot \cos(\phi) = 0$$

$$v_B = \dot{x}_B \cdot \cos(\theta) + \dot{y}_B \cdot \sin(\theta)$$

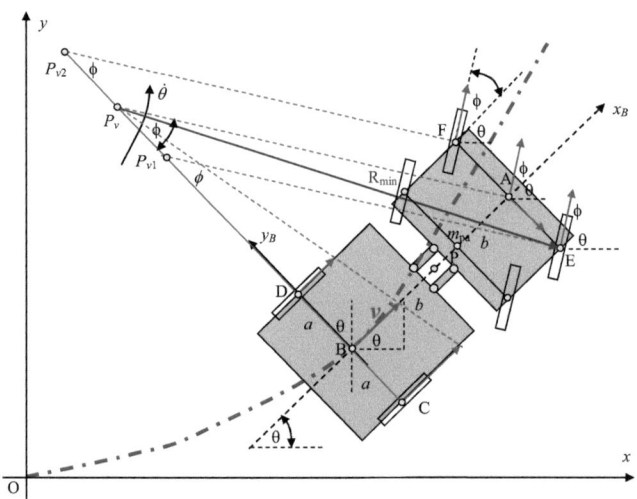

Fig. 4-2 Positional analysis and kinematic model for HUMI

---

[13] Similar version of formulas 4.1 - 4.10 can be found at (Shala et al., 2011) and (X.Bajrami, 2010)

From above follows the conditions of non-slipping in matrix form:

$$C(s) \cdot \dot{s} = 0, \qquad (4.5)$$

where :

$s = [x_B \ y_B \ \theta]^T$ and $\dot{s} = [\dot{x}_B \ \dot{y}_B \ \dot{\theta}]^T$ - represents general variables,

and finally :

$$C(s) = \begin{bmatrix} \sin(\theta) & -\cos(\theta) & 0 \\ \sin(\theta + \phi) & -\cos(\theta + \phi) & -2b\cos(\phi) \end{bmatrix} \text{ - represents the Pffafian's}$$

Matrix.

The solution of equation (5) actually represents the inverse kinematic of HUMI.

From Fig. 4-2 can be calculated also the following expressions:

$$v_B = v = \frac{b}{\tan(\phi)} \cdot \dot{\theta} \implies \dot{\theta} = \frac{v}{b}\tan(\phi) \ . \qquad (4.6)$$

and

$$\dot{x}_B = v \cdot \cos(\theta) \ , \ \dot{y}_B = v \cdot \sin(\theta) \qquad (4.7)$$

The example-trajectory followed from HUMI is taken to be a sinusoidal trajectory. The reason of using a sinusoidal trajectory $f(x) = 10.\sin(\frac{x}{6})$ is the difficulty of trajectory-following from mobile robots because of its pliability (Shala et al., 2011). It means, in case the HUMI follows this trajectory with high accuracy, it is expected at least the same accuracy also by another trajectories.

Another issues that should be taken in consideration are the HUMIS dimension (a, b) (see Fig. **4-2**) and the risk of possibility to excess the minimal turning cycle that can be achieved from HUMI. For example for the maximal turning angle ϕ_{max}=30° follows :

$$R_{B\min} = \overline{PvB} = \frac{2b}{\tan(30°)} = 2b \cdot \sqrt{3}$$

$$R_{\min} = \overline{P_vE} = \sqrt{(\overline{P_vB} + \overline{BC})^2 + (\overline{CE})^2} = \sqrt{(\overline{P_vB} + a)^2 + (2b)^2} = \sqrt{a^2 + 4ab\sqrt{3} + 16b^2}$$

$$R_{\min} > \sqrt{a^2 + 4ab\sqrt{3} + 16b^2} \tag{4.8}$$

For given $a = 0.325$ m and $b = 0.50$ m, then the pliability-radius of the trajectory $y(x)$ should satisfies the condition:

$$R_{B\min} > 2b \cdot \sqrt{3} = 1.732 \; [m] \text{ and } R_{\min} > \sqrt{a^2 + 4ab\sqrt{3} + 16b^2} = 2.287 \; [m]$$

From the equation (7), can be calculated the angle θ :

$$\theta = a\tan\left(\frac{\dot{y}_B}{\dot{x}_B}\right) \quad \text{... see (Fig. 4-9)} \tag{4.9}$$

and from equation (6) can be calculated angle ϕ :

$$\phi = a\tan\left(\frac{b \cdot \dot{\theta}}{v}\right) \quad \text{... see (Fig. 4-9)} \tag{4.10}$$

In Fig. 4-3 is presented the implementation of MATLAB model of inverse kinematic of HUMI.

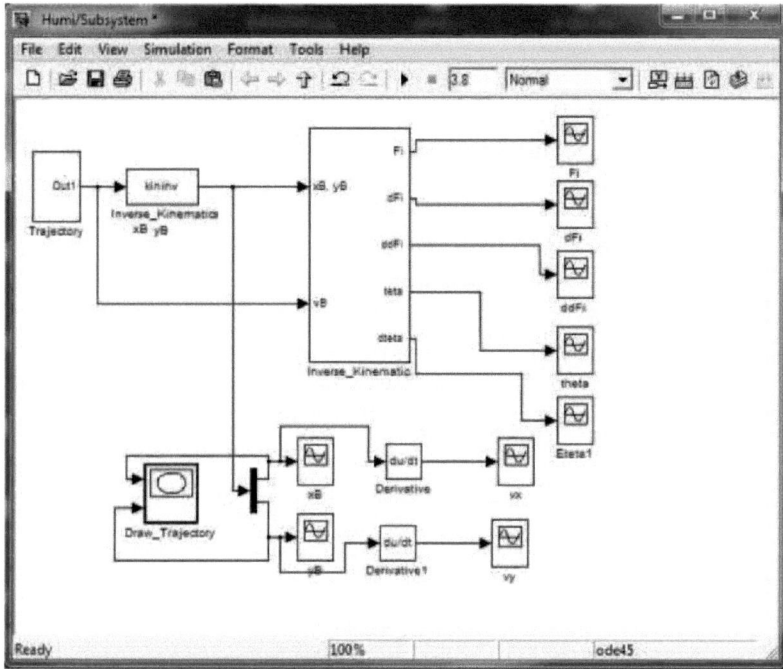

Fig. 4-3 MATLAB Model-solution for inverse kinematic of HUMI

The following Fig.s (Fig. **4-4** - Fig. 4-10) show the simulation results of HUMI model, for position of HUMI on X and Y axis during his moving as well as angle of HUMI platform and steering angle of front wheels.

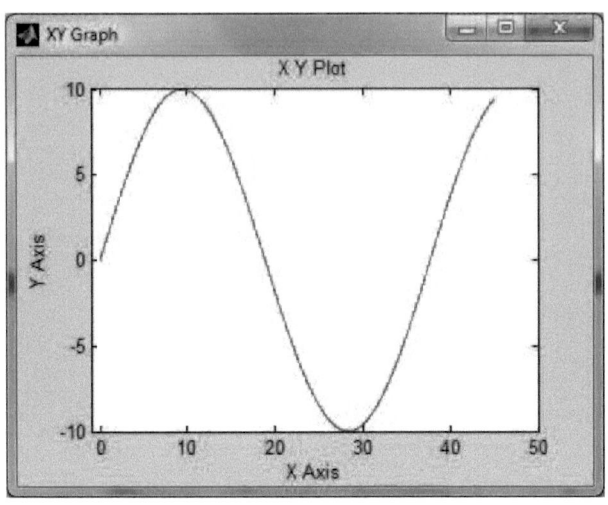

Fig. 4-4 Desired Trajectory of HUMI

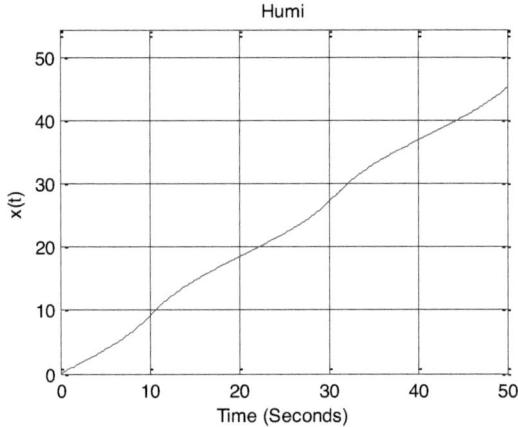

Fig. 4-5 Coordinates of point B in X axe-direction

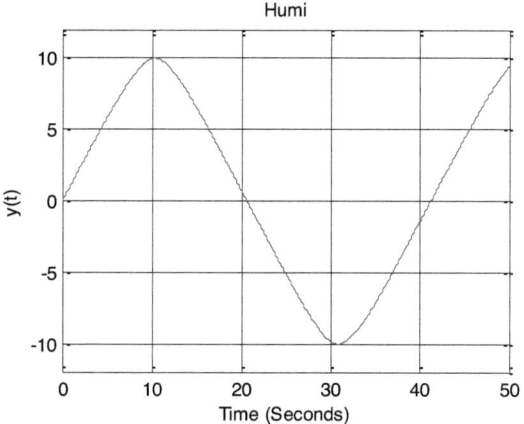

Fig. 4-6 Coordinates of point B in Y axe-direction

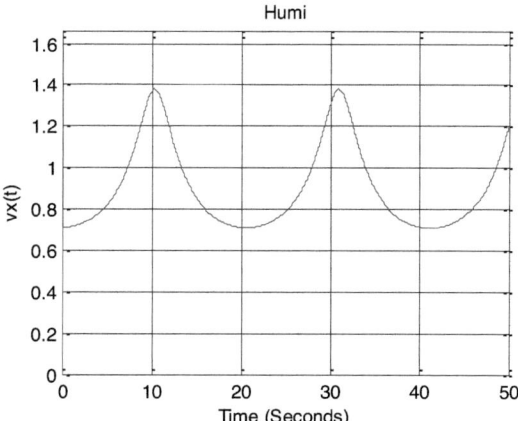

Fig. 4-7 Velocity v_x of point B

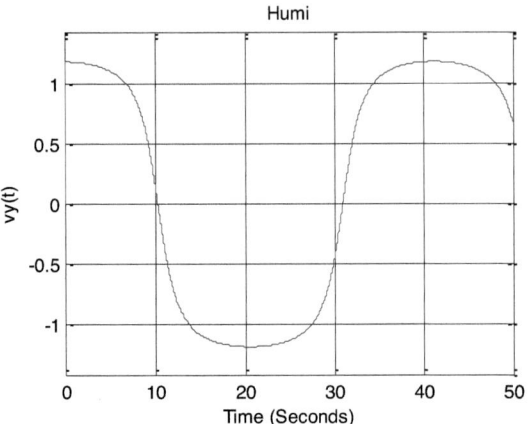

Fig. 4-8 Velocity v_y of point B

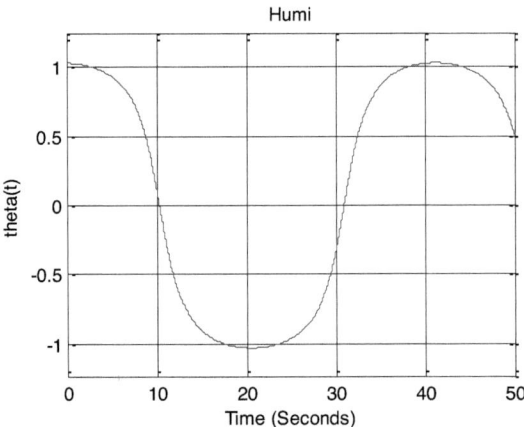

Fig. 4-9: The angle θ between HUMI and X - axe

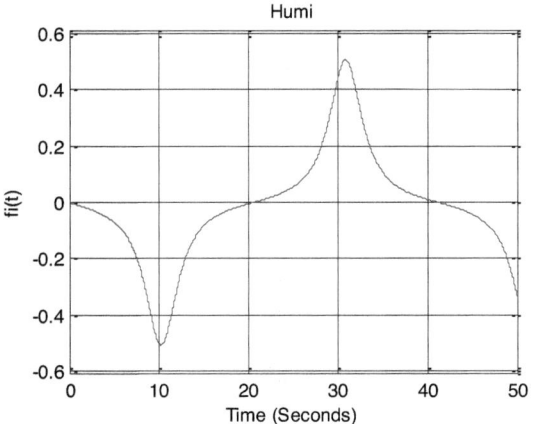

Fig. 4-10 Steering angle ϕ of front wheels of HUMI

4.1.2 Analysis of the Dynamic Model of HUMI

For finding the model of dynamic behavior of HUMI the Lagrange formulation is used:

$$\frac{d}{dt}\left(\frac{\partial L}{\partial \dot{q}}\right) - \frac{\partial L}{\partial q} = \frac{\delta A}{\delta t},$$

where: $L = E_k - E_p$ represents Lagrange function,

E_k - kinetic energy of HUMI,

E_p - potential energy of HUMI,

A - working force – for friction force,

$q = [r, \phi] \Rightarrow \dot{q} = [v, \dot{\phi}]$ - control variables equal with vehicle DOF-s..

t – time.

After the calculation of Lagrange function (kinetic and potential energy), the dynamical equations of the vehicle can be expressed in the matrix form:

$$D(q) \cdot \ddot{q} + H(q, \dot{q}) = \tau$$

where: $D(q) = \begin{bmatrix} D_{11} & D_{12} \\ D_{21} & D_{22} \end{bmatrix}$, $H(q,\dot{q}) = \begin{bmatrix} H_1 \\ H_2 \end{bmatrix}$ dhe $\tau = \begin{bmatrix} F_{sh} \\ M_d \end{bmatrix}$

F_{sh} - Nominal driving force acting on the rear axle,

M_d - Nominal torque for steering wheels.

$D(q)$ - 2x 2 matrixes, which members are in function of coordinate's q,

$H(q,\dot{q})$ - 2x1 vector, which members are in function of q and velocity $\dot{q}$

τ - 2x1 vector.

The kinetic energy of HUMI-s model can be expressed as: $E_k = E_k^{pa} + E_k^{rr}$
where :

E_k^{pa} - Kinetic energy of HUMI-s platform,

E_k^{rr} - Kinetic energy of HUMI-s wheels,

Kinetic energy of HUMI platform will be:

$$E_k^{pa} = \frac{1}{2} \cdot m_{pa} \cdot v_P^2 + \frac{1}{2} \cdot J_P \cdot \dot{\theta}^2$$

where:

m_{pa} - HUMI platform mass,

m_{per} - general mass,

$J_P = \frac{1}{2} \cdot m_{pa}(4a^2 + b_1^2)$ - is moment of inertia for HUMI-s platform

$2a$ - width of HUMI-s platform

$b_1 > 2b$ - length of HUMI-s platform

The final matrix Lagrange equations which describe the movement of the robot HUMI are:

see below, equation (4.11).

$$\begin{bmatrix} m_{pa} + 12m_{rr} + m_{per}\tan^2(\phi) & m_{rr}\dfrac{R^2}{4b}\tan(\phi) \\ m_{rr}\dfrac{R^2}{4b}\tan(\phi) & m_{rr}\dfrac{R^2}{2} \end{bmatrix} \cdot \begin{bmatrix} \dot{v} \\ \ddot{\phi} \end{bmatrix} + \begin{bmatrix} \left(2m_{per}\cdot v\cdot\dot{\phi}\cdot\tan(\phi) + m_{rr}\dfrac{R^2}{4b}\cdot\dot{\phi}^2\right)\cdot(\tan^2(\phi)+1) \\ \left(m_{per}\cdot v^2\cdot\tan(\phi) + m_{rr}\dfrac{R^2}{2b}\cdot v\cdot\dot{\phi}\right)\cdot(\tan^2(\phi)+1) \end{bmatrix} = \begin{bmatrix} \tau_1 \\ \tau_2 \end{bmatrix}$$

(4.11)

where τ_1 denotes the nominal driving force acting on the rear axle and τ_2 denotes torque *for* steering wheels.

To be able to follow the sinusoidal trajectory, the robots HUMI nominal driving force $\tau_1(t)$ (acting on the rear axle CD), and its torque for steering wheels $\tau_2(t)$ (nominally to front wheels E, F) are given at equation (4.11). The diagrams for different time values are given at Fig. 4-11 and Fig. 4-12.

Fig. 4-11 and Fig. 4-12 show :

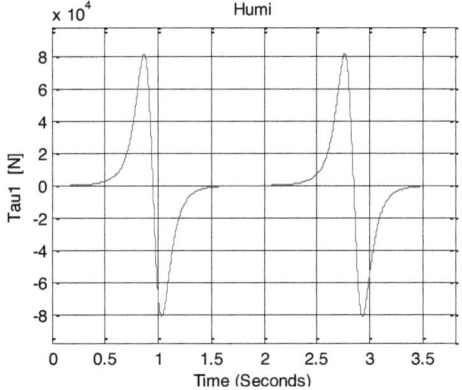

Fig. 4-11 The nominal driving force acting on the rear axle

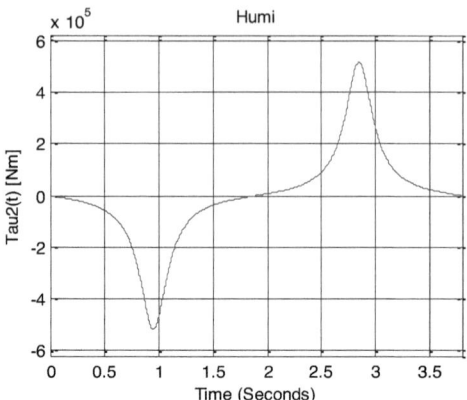

Fig. 4-12 The torque for steering wheels.

In order to reduce the order of dynamic mathematical model of robot HUMI (see equation 4.11), there will be switching to the new variables of order one.

$$x(t) = \begin{bmatrix} x_1 \\ x_2 \end{bmatrix} = \begin{bmatrix} x_1 \\ \dot{x}_1 \end{bmatrix} = \begin{bmatrix} q \\ \dot{q} \end{bmatrix} = \begin{bmatrix} [r \quad \phi]^T \\ [v \quad \dot{\phi}]^T \end{bmatrix} \qquad (4.12)$$

4.1.3 HUMI-s new controller design in FLC (Fuzzy Logic Control)

In case of path following from robot HUMI, in the open-loop feedback control strategy the velocity and torque are in function of the calculated path and its initial start and end-position. The problem here is the absence of an error-model due to the beforehand calculations. It means there exist no possibility for error compensation.

In the closed loop strategies the velocity and torque are functions of the actual state of the system and not only of the initial and end points. Therefore disturbances and errors causing deviations from the desired path are compensated by the use of the inputs. There are several available closed loop control systems, like, proportional control (P), proportional integral control (PI), proportional integral derivative control (PID), fuzzy logic control (FLC) etc.

In this work the Fuzzy logic control is selected for implementation, as so far it is known, for a highly nonlinear robot model, the fuzzy logic is one of the easiest approaches for implementation, and is well suited to low-cost implementations based on cheap sensors.

As inputs to the Fuzzy Logic Controller are taken: distance error (equation 4.13) and angle error (equation 4.14).

From the trajectory can be obtained the following distance error:

$$d_{err} = \pm\sqrt{(y_{desired} - y)^2 + (x_{desired} - x)^2} \qquad (4.13)$$

where:

$y = f(x)$ - represents the actual trajectory, $y_{desired} = f(x_{desired})$ - represents the desired trajectory of the HUMI at point B. For $sign(y_{desired} - y) = 1$ follows $y_{desired} > y$ and $d_{err} > 0$. For $sign(y_{desired} - y) = -1$ follows $y_{desired} < y$ and $d_{err} < 0$.

From the trajectory can be obtained the following HUMI body rotation error (θ):

$$e = \theta_d - \theta \qquad (4.14)$$

where: θ_d - represents the desired angle of the HUMI to the x axis, and θ - represents the actual measured angle.

Below in Fig. 4-13 the determining of rules between inputs and outputs for Fuzzy logic is given.

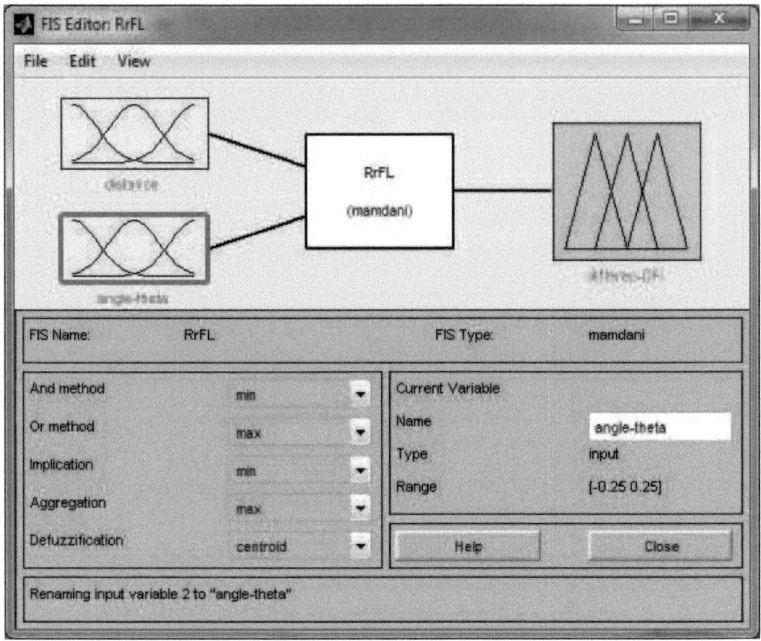

Fig. 4-13 The Fuzzy rules between Inputs and Outputs

The simulation results show that the applying of FLC (Fuzzy logic controller) reduces significantly the errors of front wheels steering angle (ϕ) as well as error of angle between HUMI platform and X axe (θ).

For front wheels steering angle (ϕ) the error after FLC is reduced from $\left[\pm 1.5 \times 10^{-3}\right]$ to $\left[\pm 4 \times 10^{-3}\right]$. (see Fig. **4-14** and Fig. **4-16**)

For angle between HUMI platform and X axe (θ) the error after FLC is reduced from $\left[\pm 1.5 \times 10^{-3}\right]$ to $\left[\pm 6 \times 10^{-3}\right]$. (see Fig. **4-15** and Fig. **4-17**)

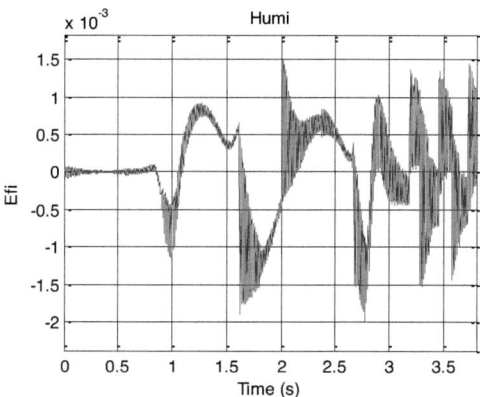

Fig. 4-14 Error of front wheels steering angle after applying Fuzzy

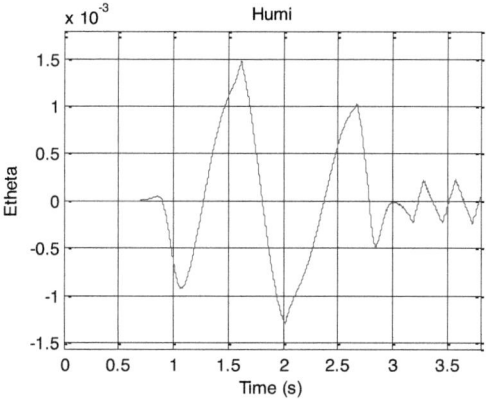

Fig. 4-15 Error for angle theta between HUMI platform and X-axe after applying Fuzzy

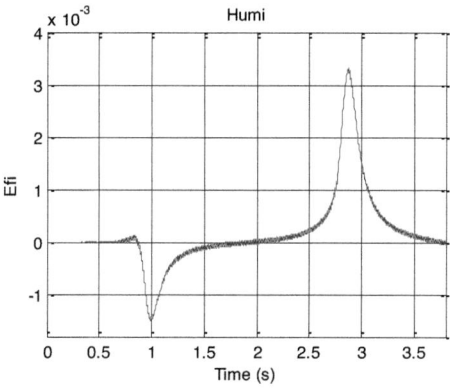

Fig. 4-16 Error of front wheels steering angle without applying Fuzzy logic

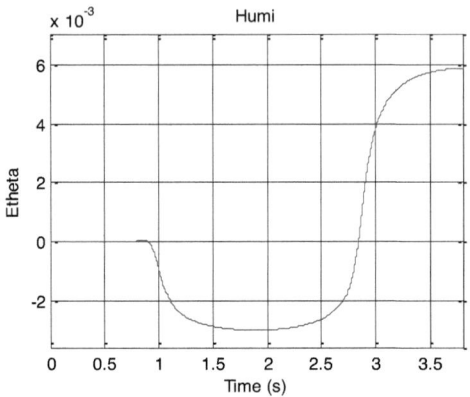

Fig. 4-17 Error for angle theta between HUMI platform and X axe without applying Fuzzy

The complete implementation of the FLC for robot HUMI is introduced into Chapter 5 (Software Implementation and Simulation results).

4.2 Humanoid Robot Archie

At the Institute of Handling Robotics and Technology (IHRT) of Vienna University of Technology a humanoid robot named Archie is designed and developed (Baltes,Byagowi,Kopacek, 2010).

The main purpose was implementing of a biped robot platform which seems to be and act like a human. The platform is thought to be of human size, low cost, modular and to perform a natural looking. This Humanoid Robot has a height of 150 cm, weight of 20 kg, with 60 min operating time by walking speed 0.5 m / s and with 29 degrees of freedom. (see Fig. 4-18)
.

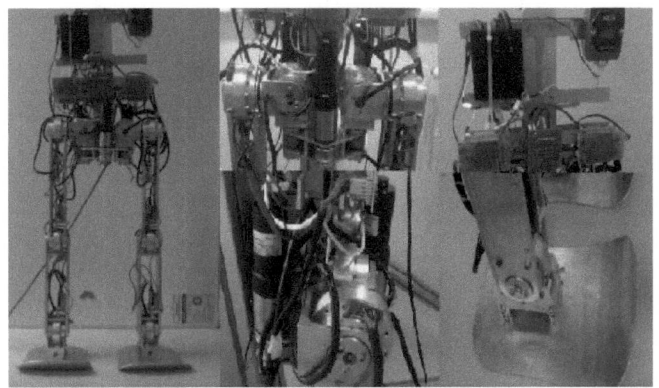

Fig. 4-18 Humanoid Robot Archie at IHRT

The mechanical analysis of Archie and its controlling concept can be found at (Baltes,Kopacek,, 2010) . The main focus of the work in this section is the kinematic model for the reduced (seven link) Archie, as well as the motion equations for the seven links. Further a new PID controller is implemented in order to reduce the errors of angles.

4.2.1 Kinematic model of biped robot Archie

In this section [14], the kinematic model of a "reduced" biped robot with seven links is presented. This "reduced" biped robot consists of seven rigid links which are connected by six purely rotational joints, each having one degree of rotational freedom. The biped robot is analyzed and modeled only for walking on a horizontal flat surface. The model in this work consists of seven rigid links: see $\theta_1...\theta_7$ in Fig. 4-19 (a).

---

[14] Some part and equations found on this section were also part of the projectwork of X.Bajrami at Institute of Mechatronic at TU-Wien

These links are connected to one another by six rotational and frictionless motors. There are two motors at the hip, two motors at the knees, and two motors at the ankles. For the stability analysis of the bipedal model, the *Zero Moment Point concept (ZMP)* will be used. To simplify the dynamic analysis, the following assumptions are made:

- The right and left sides of biped robot are symmetric.
- The biped robot is constrained in the sagittal plane (see Fig. 4-19 (b))
- To prevent slippage of the biped robot, there is sufficient friction between the soles and horizontal surface.

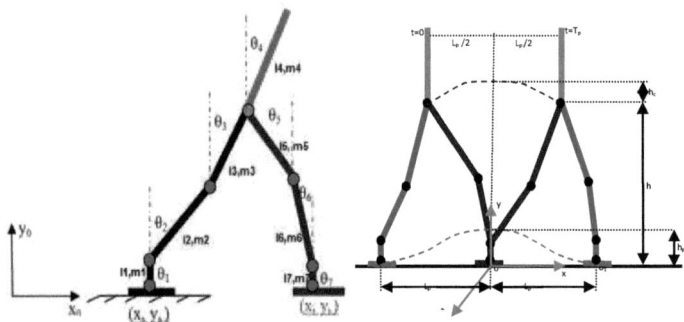

Fig. 4-19 a) Seven link biped model of Archie b) Half cycle of walking in sagittal plane

At Fig. 4-19 (b), the blue-curve is path done from the heap, the red-curve is path done from the floating leg. Further the following description for Fig. 4-19 (b) can be found:

- T_p = *time spent for each step,*
- L_p = *step length,*
- h = *hip lengtht,*
- h_c = *hip ripple,*
- h_p = *maximum height of the floating leg.*

The walking planning is implemented in the frontal plane. In the Fig. 4-19 it can be observed the trajectory of the hip and of leg float.

The associated model parameters are as follows:

- mi = Mass of link of i
- Ii = Moment of inertia of link i
- li = length of link i
- di = Distance between center of mass and the lower joint of link i
- θi = Angle of link i

The fixed coordinate x_0 O_0 y_0 is defined as the reference coordinate system where two following points are measured:

- (xs, ys) = The position of the point of support
- (xt, yt) = The position of the tip of the swing leg

The relation between these two points is as follows:

$$\left. \begin{array}{l} x_t = x_s + l_1 \cdot \sin\theta_1 + l_2 \cdot \sin\theta_2 + l_3 \cdot \sin\theta_3 + l_4 \cdot \sin\theta_4 + l_5 \cdot \sin\theta_5 + l_6 \cdot \sin\theta_6 + l_7 \cdot \sin\theta_7 \\ y_t = y_s + l_1 \cdot \cos\theta_1 + l_2 \cdot \cos\theta_2 + l_3 \cdot \cos\theta_3 + l_4 \cdot \cos\theta_4 + l_5 \cdot \cos\theta_5 + l_6 \cdot \cos\theta_6 + l_7 \cdot \cos\theta_7 \end{array} \right\}$$

(4.15)

4.2.2 Equations of Motion

The biped walking locomotion is constrained in the sagittal plane and it does not have any lateral movement. One complete gait cycle consists of two steps that include *SSP* and *DSP* as follows:

- *Both legs are in contact with the ground (Double Support Phase "DSP")*

- *Right leg swings while left leg supports the biped (Single Support Phase "SSP")*
- *Both legs are in contact with the ground (Double Support Phase "DSP"))*
- *Left leg swings while right leg supports the biped(Single Support Phase "SSP")*
- *Both legs are in contact with the ground (Double Support Phase "DSP")*

Here, the position of the ZMP (Zero Moment Point) is computed by finding of the point (x,y,z) where the total torque is zero. Since we are interested for the sagittal plane, it is assumed that y=0. (see Fig. 4-20)

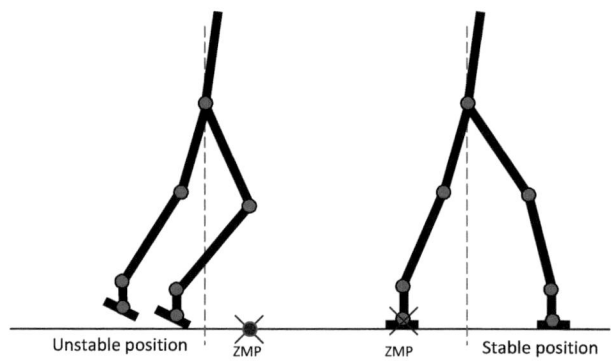

Fig. 4-20 Dynamic walking of Archie

4.2.3 Dynamic modeling in SSP model

In this work, to solve the Dynamic Modeling for Archie, the Lagrange equations are used:

$$\frac{d}{dt}(\frac{\partial L}{\partial \dot{q}}) - \frac{\partial L}{\partial q} = \frac{\partial A}{\partial t} \qquad (4.16)$$

where:

$L = E_k - E_p$

E_K : Kinetic Energy,

E_p : Potential Energy,

q : Generalized coordinate.

The kinetic and potential energy for each coordinate are given by following equations:

$$E_k = \frac{1}{2} \cdot m \cdot v_c^2 + \frac{1}{2} \cdot I \cdot \dot{\theta}^2 \qquad (4.17)$$

$$E_p = m \cdot g \cdot y_c \qquad (4.18)$$

Because the biped robot has seven links, the total kinetic and potential energy are given as follows:

$$\left. \begin{array}{l} E_k = \sum_{i=1}^{7} \cdot E_{ki} \\ E_k = E_{k1} + E_{k2} + E_{k3} + E_{k4} + E_{k5} + E_{k6} + E_{k7} \end{array} \right\} \qquad (4.19)$$

$$\left. \begin{array}{l} E_p = \sum_{i=1}^{7} \cdot E_{pi} \\ E_p = E_{p1} + E_{p2} + E_{p3} + E_{p4} + E_{p5} + E_{p6} + E_{p7} \end{array} \right\} \qquad (4.20)$$

where:

- g : Gravitational acceleration = 9.81 *m/s2*
- v_c : Velocity of center of mass of a link *(m/s)*
- y_c : Vertical coordinate of center of mass of a link *(m)*
- $\dot{\theta}$: Angular velocity of a link *(rad/s)*

Kinematic relationship between joint angles θ_i, y_{ci}, x_{ci}, and v_{ci} are given by:

$$\begin{aligned}
\begin{bmatrix} x_{c1} \\ y_{c1} \end{bmatrix} &= \begin{bmatrix} d_1 \cdot \sin\theta_1 \\ d_1 \cdot \cos\theta_1 \end{bmatrix} \\
\begin{bmatrix} x_{c2} \\ y_{c2} \end{bmatrix} &= \begin{bmatrix} l_1 \cdot \sin\theta_1 + d_2 \cdot \sin\theta_2 \\ l_1 \cdot \cos\theta_1 + d_2 \cdot \cos\theta_2 \end{bmatrix} \\
\begin{bmatrix} x_{c3} \\ y_{c3} \end{bmatrix} &= \begin{bmatrix} l_1 \cdot \sin\theta_1 + l_2 \cdot \sin\theta_2 + d_3 \cdot \sin\theta_3 \\ l_1 \cdot \cos\theta_1 + l_2 \cdot \cos\theta_2 + d_3 \cdot \cos\theta_3 \end{bmatrix} \\
\begin{bmatrix} x_{c4} \\ y_{c4} \end{bmatrix} &= \begin{bmatrix} l_1 \cdot \sin\theta_1 + l_2 \cdot \sin\theta_2 + l_3 \cdot \sin\theta_3 + d_4 \cdot \sin\theta_4 \\ l_1 \cdot \cos\theta_1 + l_2 \cdot \cos\theta_2 + d_3 \cdot \cos\theta_3 + d_4 \cdot soc\theta_4 \end{bmatrix} \\
\begin{bmatrix} x_{c5} \\ y_{c5} \end{bmatrix} &= \begin{bmatrix} l_1 \cdot \sin\theta_1 + l_2 \cdot \sin\theta_2 + l_3 \cdot \sin\theta_3 - (l_5 - d_5) \cdot \sin\theta_5 \\ l_1 \cdot \cos\theta_1 + l_2 \cdot \cos\theta_2 + d_3 \cdot \cos\theta_3 - (l_5 - d_5) \cdot \cos\theta_5 \end{bmatrix} \\
\begin{bmatrix} x_{c6} \\ y_{c6} \end{bmatrix} &= \begin{bmatrix} l_1 \cdot \sin\theta_1 + l_2 \cdot \sin\theta_2 + l_3 \cdot \sin\theta_3 - l_5 \cdot \sin\theta_5 - (l_6 - d_6) \cdot \sin\theta_6 \\ l_1 \cdot \cos\theta_1 + l_2 \cdot \cos\theta_2 + d_3 \cdot \cos\theta_3 - l_5 \cdot \cos\theta_5 - (l_6 - d_6) \cdot \cos\theta_6 \end{bmatrix} \\
\begin{bmatrix} x_{c7} \\ y_{c7} \end{bmatrix} &= \begin{bmatrix} l_1 \cdot \sin\theta_1 + l_2 \cdot \sin\theta_2 + l_3 \cdot \sin\theta_3 - l_5 \cdot \sin\theta_5 - l_6 \cdot \sin\theta_6 - (l_7 - d_7) \cdot \sin\theta_7 \\ l_1 \cdot \cos\theta_1 + l_2 \cdot \cos\theta_2 + d_3 \cdot \cos\theta_3 - l_5 \cdot \cos\theta_5 - l_6 \cdot \cos\theta_6 - (l_7 - d_7) \cdot \cos\theta_7 \end{bmatrix}
\end{aligned}$$
(4.21)

After derivation of equations 21 in respect to coordinates θ_i for $i \in \{1,2...7\}$ and their substitutions into equations (17 and 18) the kinetic and potential energy is obtained. Further the equations (17, 18) will be substitutes into second kind Lagrange equations (16), and after partial derivation in respect the time component, the n-link equations of motion in Matrix- Lagrange Form will be obtained (see equation 22).

$$\begin{bmatrix} M_{11} & M_{12} & \cdots & M_{1n} \\ M_{21} & M_{22} & \cdots & M_{2n} \\ \vdots & \vdots & \vdots & \vdots \\ M_{n1} & M_{n2} & \cdots & M_{nn} \end{bmatrix} \cdot \begin{bmatrix} \ddot{\theta}_1 \\ \ddot{\theta}_2 \\ \vdots \\ \ddot{\theta}_n \end{bmatrix} + \begin{bmatrix} \varepsilon_1 \\ \varepsilon_2 \\ \vdots \\ \varepsilon_n \end{bmatrix} + \begin{bmatrix} H_1 \\ H_2 \\ \vdots \\ H_n \end{bmatrix} = \begin{bmatrix} \tau_1 \\ \tau_2 \\ \vdots \\ \tau_n \end{bmatrix} \qquad (4.22)$$

where

- M_{ij} are the inertial terms,
- ε_i are the gravitational terms,
- H_i are the Carioles and the centrifugal terms,

- τ_i are the input torques.

Finally a proportional integrative derivation controller (PID) is implemented in Matlab/Simulink in order to regulate angles during trajectory following from Archie.

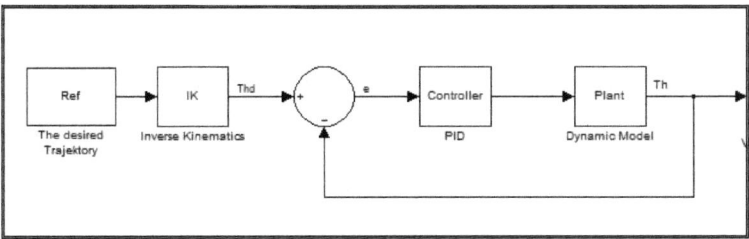

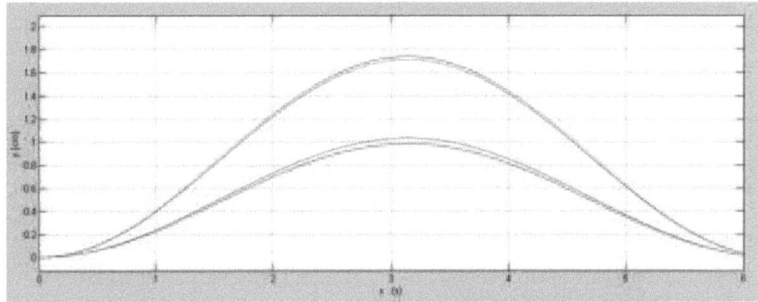

Fig. 4-21 Simulation results after applying of PID for Archie

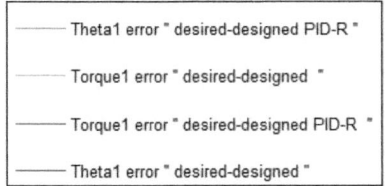

Simulation results obtained from

Fig. **4-21** show that after applying of PID regulator, although there are some errors-reduce, however the results are actually poor.

Chapter 5

5 Software Implementation and Simulation Results

In this work, a simulation-application is developed. It is a windows-platform application implemented in C# (Visual Studio 2010) and Matlab 2009 b. The main part of application, are:

- *map-creation* : according to environment-size, cell-size, number of cells or robot-size
- *path-find-methods* : three different approaches implemented:
    - Hypertree Decomposition
    - Geometric Intersection
    - Dijkstra
- *Error-Estimation methods* : Several error estimation methods during the localization
    - Extended Kalman Filter,
    - Neural Network
    - Fuzzy Logic
- *PID-Controller* : for Humanoid Robot Archie
- *Simulation-of-path following*: A simulation of path following from mobile robot, based on above created-map, path-find-method and error estimation approach.

The whole application has approximately 3000 rows of code.

5.1 Map-Creation

The user can simulate a map-grid according to environment-size, cell-size, number of cells or robot-size. On Create Map button click, the desired environment will be created automatically on the panMap, and a visualized grid-map with several data like height and width of environment, cells size, number of cells etc will be presented in application-window (see Fig. **5-1**).

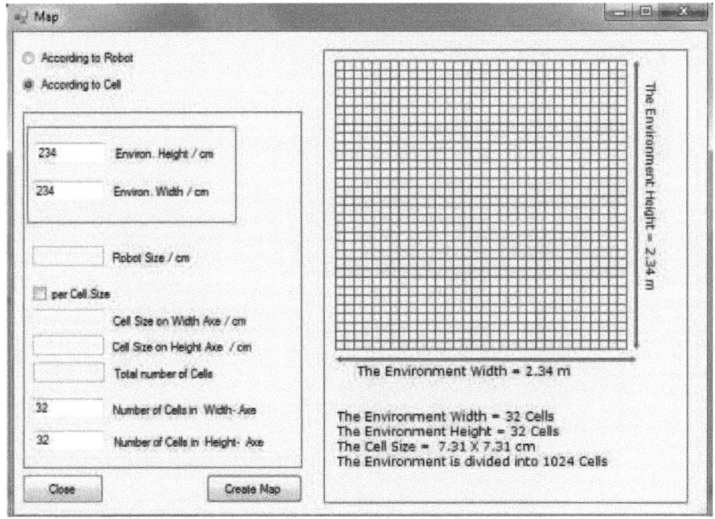

Fig. 5-1 Simulation of Map Creation

The code of implementation for map-creation is written mainly on the *class Map.cs* . The piece of the code can be found in *Appendix A*.

The simulated grid-map further can be filled with obstacles and used further on the main window.

5.2 Path-Simulating

On the main window, clicking on button RefreshMap, the simulated grid-map will be imported and by selecting on HypertreeDecomposition, GeometricIntersection or Dijkstra algorithm, a optimal or near optimal path will be generated and on the map visualized. (see Fig. **5-2**). Inserting as well as deleting on the map of Obstacles, StartPosition, or TargetPosition can be done first by selecting of respect methods on Obstacles panMap, and then by mouse-moving on the map. (see Fig. 5-2) .

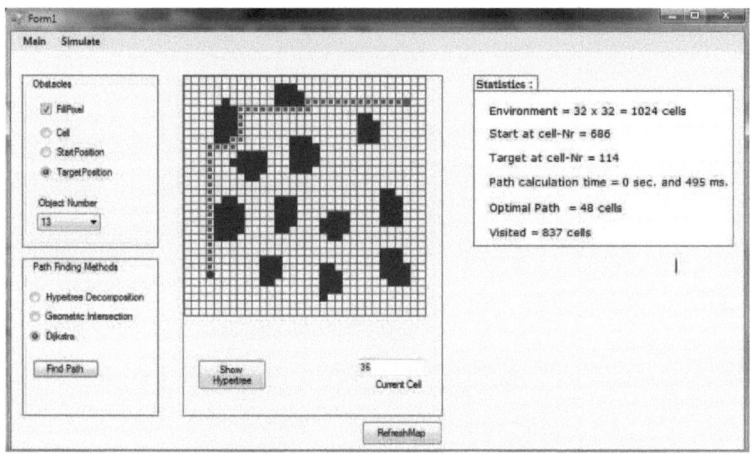

Fig. 5-2 Main Window of Simulation application

By button-click ShowHypertree , can be visualized the actual Hypertree Decomposition, obtained from heuristic. (see Fig. 5-3) .

The statistics on the right side (see Fig. Fig. **5-2**) show the environment data, start and target positions, path calculation time, the optimal path found as well as all visited cells during the path calculation.

In case the statistics for all three algorithms are required, then clicking in the Menu Item Main->Statistics , the statistical data for different algorithms can be obtained. (see Fig. 5-4).

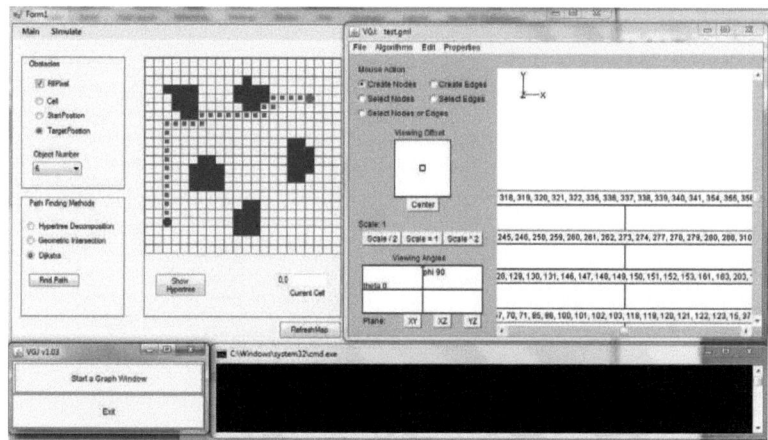

Fig. 5-3 Simulating of Hypertree Decomposition

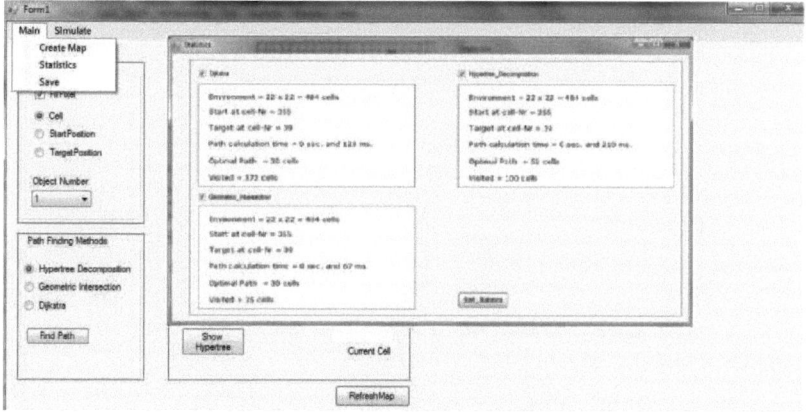

Fig. 5-4 Statistics of all approache

Below (Table 5.1) can be found the test results after simulating of the path-finding for several algorithms:

Map-Dimens	Nr.of .Obstacles	Start Cell Posit.	Target Cell Posit.	Dijk.Alg	HypDecomAlg	IntersAlg
22x22	8	339	60	133 ms (calc.time) 24 cells (found. Path) 387 visited cells	2 ms 25 cells 50 cells	3 ms 24 cells 26 cells
22x22	8	382	43	124 32 16	3 33 66	8 32 34
22x22	8	321	37	123 32 405	110 34 71	61 35 46
22x22	8	373	76	117 35 405	3 38 76	12 43 47
22x22	8	373	73	117 32 338	3 35 70	11 38 42
15x15	5	167	4	24 15 167	1 15 31	5 15 16
15x15	5	167	26	24 18 185	2 21 40	8 20 26
15x15	5	167	17	25 16 175	1 16 33	3 16 16
34x36	14	913	412	575 33 767	614 44 81	70 58 62
34x36	14	913	38	656 34 795	662 45 85	27 34 36
34x36	14	995	336	717	-	26

				30		30
				692		32
32x22	7	504	88	268	250	63
				33	45	33
				571	62	38
32x22	7	540	62	270	269	13
				37	57	37
				620	85	39
32x22	7	552	54	268	250	11
				21	29	29
				438	55	32

Table 5.1 The simulation results for different implemented algorithms

Table 5.1 shows that with the new implemented algorithms calculation time is improved up to 20-time, during the optimality of path is decreased only 5-10%. Because the new implemented algorithms in this work are heuristic algorithms, the results strongly depend from the implemented heuristic, and can be varied from case to case. In general, the results are significant better than obtained from exact algorithm of Dijkstra.

5.3 Path Simulation for HUMI

In this section a overview about MATLAB solution for HUMI is given. In Fig. 5-5 a complete scheme in Matlab is given.

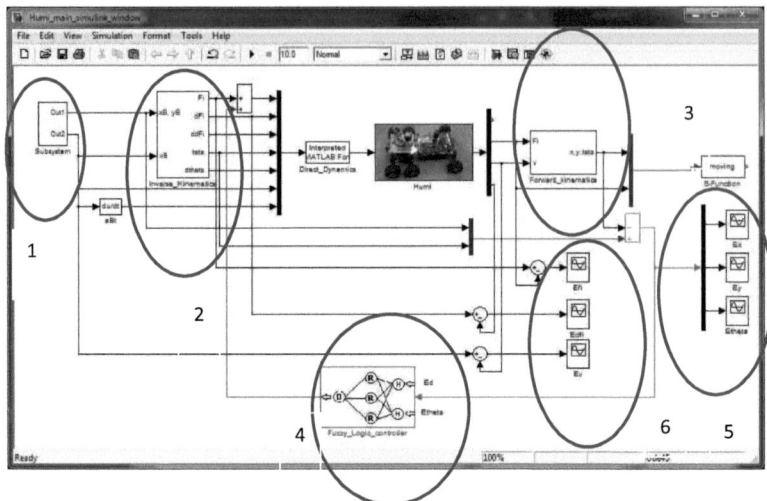

Fig. 5-5 MATLAB Model solution for HUMI

The 5 main components (see Fig. 5-5 red colors) are described:

- Componet 1: Component Subsystem calling some of data (velocity and of inverse kinematic) of HUMI given in M-File. (see Fig. 5-6)

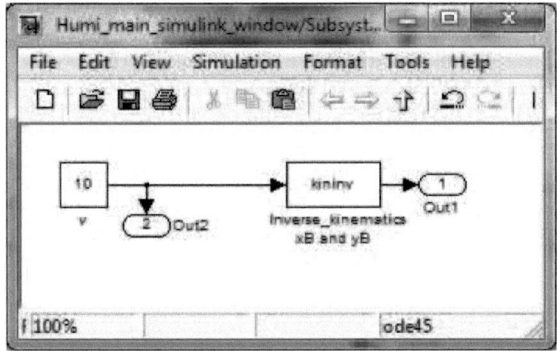

Fig. 5-6 Subsystem for Inverse Kinematic – M.File

- Component 2: Subsystem calling other data (angle theta for platform, velocity angle dtheta, etc see fig) of inverse kinematic of HUMI given as blocks. (see Fig. 5-7)

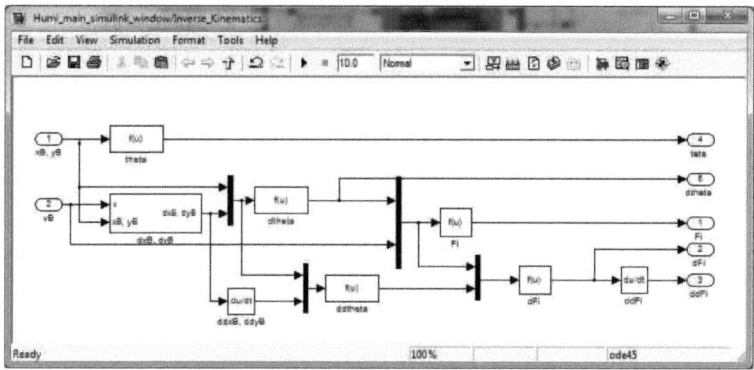

Fig. 5-7 Inverse Kinemtaic of HUMI - other data

- Component 3: Subsystem for forward kinematic. As inputs are given theta (see Fig. 5-8), velocity of rear wheel axis at point B (see fig. 4-2). As output is coordinates of platform (x,y,theta).

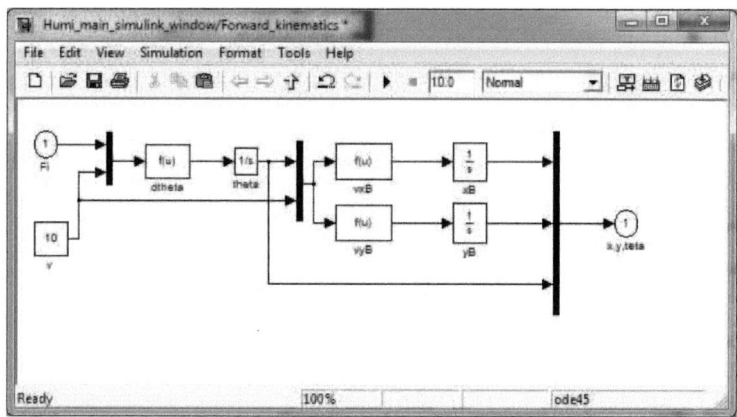

Fig. 5-8 Forward Kinematic

- Component 4: Implementation of Fuzzy Logic in respect to distance-error Ed and angle-error Etheta.

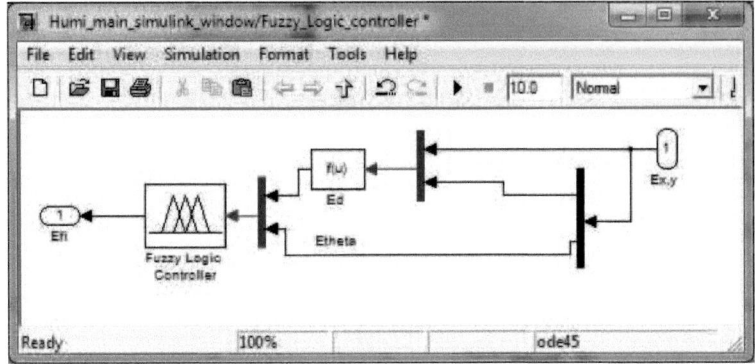

Fig. 5-9 Fuzzy Logic

Components 5 and 6: are the results obtained from HUMI without Fuzzy and with Fuzzy logic.

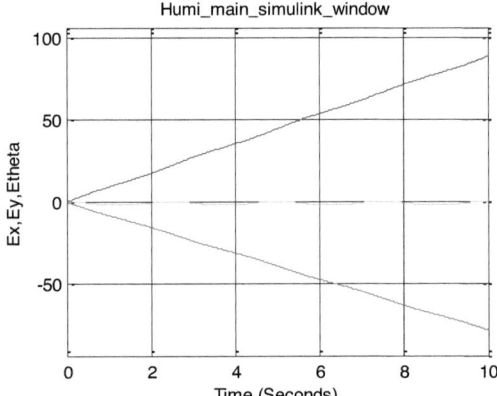

Fig. 5-10 Errors of position and angle for HUMI-platform

Fig. 5-10 represents the errors of position (Ex and Ey) and errors of angle (Etheta) for HUMI, in case no Fuzzy Logic is applied.
- Green curve represents the angle-error [degree] (Etheta) for HUMI platform

- Blue curve represents the error of platform Ex in X-axe direction for HUMI
- Red curve represents the error of platform Ey in Y-axe direction for HUMI

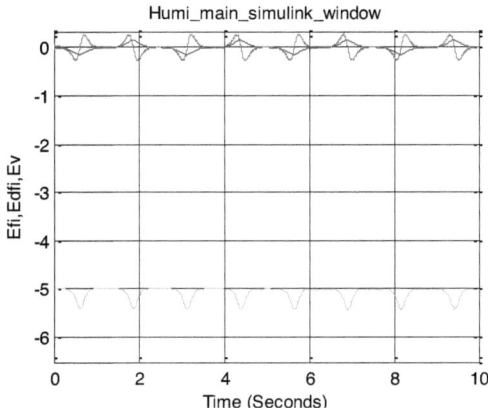

Fig. 5-11 Error of front wheels steering angle and velocity for HUMI-platform

Fig. 5-11 represents the error of velocity (Ev), Error of front wheels steering angle (Efi), and error of velocity of angle (Edfi) for HUMI, in case no Fuzzy Logic is applied. The simulation
- Green curve represents the Ev for HUMI platform
- Blue curve represents the Efi for HUMI platform
- Red curve represents the Edfi for HUMI platform

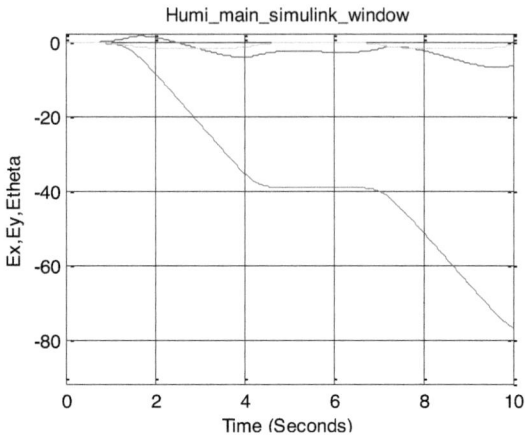

Fig. 5-12 Errors of position and angle for HUMI-platform with FLC

Fig. 5-12 represents the errors of position (Ex and Ey) and errors of angle (Etheta) for HUMI, in case the Fuzzy Logic is applied.
- Green curve represents the angle-error [degree] (Etheta) for HUMI platform
- Blue curve represents the error of platform Ex in X-axe direction for HUMI
- Red curve represents the error of platform Ey in Y-axe direction for HUMI

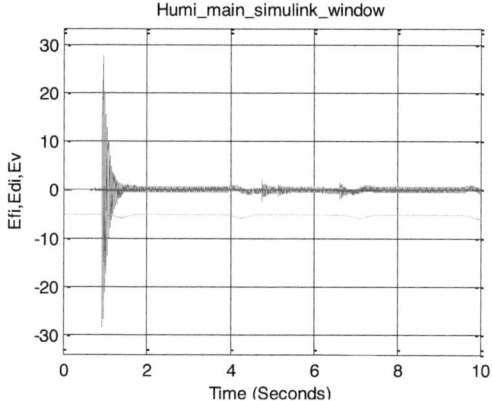

Fig. 5-13 Error of front wheels steering angle and velocity for HUMI-platform with FLC

Fig. 5-13 represents the error of velocity (Ev), Error of front wheels steering angle (Efi), and error of velocity of angle (Edfi) for HUMI, in case the Fuzzy Logic is applied. The simulation

- Green curve represents the Ev for HUMI platform
- Blue curve represents the Efi for HUMI platform
- Red curve represents the Edfi for HUMI platform

Chapter 6

6 Summary and Outlook

In this work new heuristics approaches based on hypertree-decomposition and geometrical intersection between robot and obstacles are implemented and tested. These new approaches calculate an optimal or nearly optimal path from robot to the target object and are implemented in C# (Visual Studio 2010). Further, a whole system beginning from localization and map building, path-planning up to motion control is proposed and in particular also implemented. The state of the art approaches like EKF, Neuronal Network, Fuzzy Logic and PID Controller are for this purpose also implemented (Matlab) and tested.

The new heuristic algorithms based on hypertree decomposition and geometrical intersections are tested, and have show very good results in respect to Dijkstra Algorithm. The main benefit of the new approaches is reducing of time-calculation, where in some cases calculation time is improved up to 20-time, during the optimality of path is decreased only 5-10%. It should be noted that the results strongly depends from the implemented heuristic and the next task can be the improving of the heuristic methods in order to obtain even better results.

In case of multi agent system robots, where more robots moves on the same area these approaches can be ideal solution because of the improving of time-calculation. For every new path following for robot, Dijkstra algorithm for example recalculates and operates on graph structure, during the hypertree decomposition recalculates only on tree-structure. For thousands such recalculations, significant time and other source benefits can be achieved.

Also the implementation of Neuronal Network, Fuzzy Logic and PID controller has shown the improving of estimated positions of robot during its localization and motion.

6.1 Future Work

In this work, there are several improvements possible that can be applied in order to achieve a better performance. In this section some of these improvements and suggestions are presented for the future work. Especially there are suggestions for the improvements of the heuristic methods.

In case of Hypertree Decomposition, is already noted that the heuristic used for the path recognizing has a strong impact on the quality of optimality. In this work, the path is a set of neighbor's grid-map cells, where each one has a number and position. The partition of grid-map with cells, the dividing of grid-map on several area and after the using of different algorithms for hypertree decomposition (hMetis, Fiduccia, etc) strongly impact the quality of decomposition. The better decomposition heuristic, leads to better choosing of neighbor cells. The problem is that in such heuristics sometimes only after lot of experiments can be decided for a better solution (and not theoreticely). The next task for the heuristic can be experimenting with different decomposition algorithms, and different possibility of grid-map dividing into areas.

In case of Geometrical Intersection approach, significant improving can be done by the obstacle avoiding heuristic. Decision for "left" or "right" avoiding strategy in case the robot is in front of an object, in this work is done as local-decision. The problem is that sometimes although the local decision seems to achieve a better local solution, in general decrease the optimality. The next task for this heuristic can be using of histograms to propagate the lower density of obstacles.

Also in case of simulation-application, including of vision system in order to obtain the map-grid automatically from robot environment can be implemented as next task. After the direct connection between software and real robot, the application like EKF, Neural Network, Fuzzy and other application can be tested in real time.

Bibliography

Niemueller,T. and Widyadharma,S. (2003). *Artificial Intelligence - An Introduction to Robotics.* RWTH , Achen.

A.Dermaku, X. (2013). Two new heuristic approaches for optimal path calculation on occcupancy grid map. *e & i Elektrotechnik und Informationstechnik* , 130 (2), 5460.

Baltes,J.,Byagowi,A.,Kopacek.P. (2010). A teen Sized Humanoid Robot - Archie. *IJAA*, (pp. 60-71).

Bulitko, V., Sturtevant,N., Jieshan, L., Timothy,Y. (2007). Graph Abstraction in Real-Time Heuristic Search. *Journal of Artificial Intelligence Research 30* , 51-100.

Burgard,W., Fox,D., Cremers,A. (1997). Fast Grid-Based Position Tracking for Mobile Robots. *in Procedings of 21th German Conference on Artificial Intelligence (KI97).* Freiburg Germany: Springer-Verlag.

Capek, K. (1920). Rossum's Universal Robots. http://-----.

Corke, P. (2011). *Robotics, Vision and Control.* Heidelberg: Springer-Verlag Berlin.

Dermaku, A. (2007). Generalized Hypertree Decomposition based on Hypergraph Partitioning. *Master Thesis - TU Wien.* Vienna.

Dermaku,A., Ganzow,T., Gottlob,G., McMahan,B., Musliu,N., Samer,M. (2007). Heuristic Methods for Hypertree Decomposition. *Proceedings of the 7th Mexican International Conference on Artificial Intelligence MICAI'08* (pp. 1-11). Mexiko: pub.

FIRA. (n.d.). *www.fira.net*.

Gan, Q. H. (2001). Comparation of two Measurement Fusion Methods for Kalman-filter-based Multi-sensorData Fusion. (pp. 273-280). IEE Transactions on Aerospace and Electronic Systems, Vol 37, Nr1.

Gottlob,G., Leone, N., Scarcello,F. (1999). Hypertree Decomposition and tractable queries. *PODS' 99 Procedings of the Eighteenth ACM Symposium on Principle of Database Systems.*

Hart P, Nilsson N, Raphael B. (1968). A formal basis for the heuristic determination of minimum costs path. *IEEE Transactions on System Science and Cybernetics 4(2),* (pp. 100-107).

Harvey,P. and Ghose,A. (1999). Fast Hypertree Decomposition.

Huosheng, H. G. (2005). *Sensors and Data Fusion Algorithms in Mobile Robotics.* Essex University .

Ito, D. (2009). *Robot Vision: Strategies, Algorithms and Motion Planning.* New York: Nova Science Publisher, Inc.

Jeavons,P., Cohen,D., and Gysens,M. (1991). A Structural Decomposition of Hypergraphs. *Mathematics Subject Classification* .

Kensuke, H. et al. (2010). *Motion Planning for Humanoid Robots.* London: Springer-Verlag London ISBN 978-1-84996-219-3.

Kopacek, P., Wuerzl, M., Schierer, E. (2005). Vision System and Game - Strategies for Robotsoccer. *IFAC.* ilenna.

Kopacek, P. (2010). Autonomous Mobile Robots. *International Journal Automation Austria* .

Kopacek, P. (2011). Cost Oriented Humanoid Robots. *18th IFAC World Congress.* Milano: IFAC.

Korf, R. (1990). Real-time heuristic search. *Artificial Intelligence, 42(2-3)*, (pp. 189-211).

Latombe, J. -C. (1991). Robot Motion Planning. Norwood MA, Kluwer Academic Publishers.

LaValle, M. (2006). *Planning Algorithms.* Cambridge University Presse.

Lazanas,A.and Latombe, J.C. (1992). Landmark-based robot navigation. *In Proc. 10th National Conference on Artificial Intelligence (AAAI-92)* (pp. 816-822). Cambridge.

Lozano-Perez, T., Mason, M., and Taylor, R. (1983). Automatic Syntethis of Fine-Motion Strategies for Robots . *International Symposium of Robotics Research.* Breton Woods.

Maybek, P. "The Kalman Filter: An introduction to concepts". in Introduction to Autonomous Mobile Robots.

Nakamura,T., Asada M. (1995). Motion Sketch: Acquisition on Visual Motion Gided Behaviours. *Proc 14'th Intel Joint Conf. Artificial Intelligence, vol.1,* , 126-132.

Pablo Gonzales, J. (2008, April). Planning with Uncertainty in Position Using High-Resolution Maps. Pittsburgh Pensylvania: Robotics Institute, Carnegie Mellon University,.

Pablo Gonzales, J. (2008, April). Planning with Uncertainty in Position Using High-Resolution Maps. Pittsburgh Pensylvania: Robotics Institute, Carnegie Mellon University, .

Patnaik, S. (2007). *Robot Cognition and Navigation.* Springer-Verlag Berlin Heidelberg 2007.

Putz, B. (2004). Navigation of mobile and cooperative robots (in German). *Doctor Thesis* . Vienna: IHRT, Vienna University of Technology.

R.Raol, J. (2010). *Multi-Sensor Data Fusion with Matlab.* 6000 Broken Sound Parkway NW,Suite 300: Taylor and Francis Book, LLC.

Shala,A., Bajrami,Xh.,Likaj,R. (2011). Design and Simulation of an autopilot by using Fuzzy Logic Controller. Hudem Robotics and Mechanical Assitance in Humanitarian De-mining.

SIEGWART,R. and NOURBAKHSH,I.R. (2004). *Introduction to Autonomous Mobile Robots.* Cambridge, Massachusetts: A Bredford Book.

Siegwart,R. and Nourbakhsh.I.R. (2004). *Introduction to Autonomous Mobile Robots.* Cambridge, Massachusetts: A Bredford Book.

Silberbauer, L. (2008). Development of an intelligent mobile robot for landmine detection. *PhD Thesis* . Vienna: University of Technology.

Stentz, A. (1994). Optimal and eficcient path planning for partially-known environments, Robotics and Automation. *IEEE International Conference on Robotics and Automation*, (pp. 3310-3317).

Stentz, A. (1994,In Proceedings). Optimal and eficcient path planning for partially-known environments, Robotics and Automation. *IEEE International Conference on Robotics and Automation*, (pp. 3310-3317).

Stenz, A. (1995). The foccused D* Algorithm for real-time replanning . *In Proceddings of the international Joint Conference on Artificial Intelligence (IJCAI)*, (pp. 1652-1659).

Takeda, H., Latombe, J.C. (1992). Sensory uncertainty field for mobile robot navigation. *IEEE International Conference on, 2465-2472.*

Tim Niemueller and Sumedha Widyadharma. (2003). *Artificial Intelligence - An Introduction to Robotics.* RWTH , Achen.

I want morebooks!

Buy your books fast and straightforward online - at one of the world's fastest growing online book stores! Environmentally sound due to Print-on-Demand technologies.

Buy your books online at
www.get-morebooks.com

Kaufen Sie Ihre Bücher schnell und unkompliziert online – auf einer der am schnellsten wachsenden Buchhandelsplattformen weltweit!
Dank Print-On-Demand umwelt- und ressourcenschonend produziert.

Bücher schneller online kaufen
www.morebooks.de

OmniScriptum Marketing DEU GmbH
Bahnhofstr. 28
D - 66111 Saarbrücken
Telefax: +49 681 93 81 567-9

info@omniscriptum.com
www.omniscriptum.com

Printed by Books on Demand GmbH, Norderstedt / Germany